Dual-Mass Linear Vibration Silicon-Based MEMS Gyroscope

Huiliang Cao

Dual-Mass Linear Vibration Silicon-Based MEMS Gyroscope

国防工業出版社
National Defense Industry Press

Springer

Huiliang Cao
North University of China
Taiyuan, Shanxi, China

ISBN 978-981-19-9246-9 ISBN 978-981-19-9247-6 (eBook)
https://doi.org/10.1007/978-981-19-9247-6

Jointly published with National Defense Industry Press
The print edition is not for sale in China (Mainland). Customers from China (Mainland) please order the print book from: National Defense Industry Press.
ISBN of the Co-Publisher's edition: 978-7-118-12115-5

This Springer imprint is published by the registered company Springer Nature Singapore Pte Ltd.
The registered company address is: 152 Beach Road, #21-01/04 Gateway East, Singapore 189721, Singapore

Preface

Micro electro mechanical system (MEMS) gyroscope is an important application of MEMS technology in inertia type instruments. It has the advantages of small volume, light weight, low cost, mass production and good impact resistances. It has important practical value and wide application prospect in national economy and national defense and military fields. In particular, the double mass line vibration structure is the main research object of domestic and foreign scientific research institutes. Although the precision of silicon micromechanical gyroscopes has been improved significantly in recent decades, its static and dynamic performance still cannot meet the requirements of medium and high precision fields due to various factors, including the deterioration and drift of gyroscope output signal caused by machining errors: The contradiction between mechanical sensitivity of gyro structure and gyroscope bandwidth; The detection modal dynamic performance needs to be further improved; the temperature characteristics of MEMS gyro are poor. For this purpose, the author compiled this book according to the latest development, and combined it with his research and engineering practice in the field of dual-mass linear vibration silicon-based MEMS gyroscope.

The book is divided into seven chapters: Chap. 1 is "Introduction", which introduces the basic concept, characteristics and application fields of the dual-mass linear vibration silicon-based MEMS gyroscope, the research status and the research purpose and significance of the book. Chapter 2 is "Dual-Mass Linear Vibration Silicon-Based MEMS Gyroscope Structure and Working Principle" which mainly introduces the Coriolis effect, gyro dynamics equation, fully decoupled structure design and machining technology of dual-mass linear vibration silicon-based MEMS gyroscope. Chapter 3 is "Dual-Mass Linear Vibration Silicon-Based MEMS Gyroscope Structure Noise Analysis and System Model" which introduces the main noise in the gyroscope structure, establishes the ideal gyroscope mechanical system model and studies the driving closed-loop loop based on AGC technology; Chap. 4 is "Dual-Mass Linear Vibration Silicon-Based MEMS Gyroscope Quadrature Error Nulling Technology and Optimization" where the generating mechanism of dual-mass linear vibration MEMS gyroscope quadrature error and its suppression methods are analyzed. Chapter 5 is "Dual-Mass Linear Vibration Silicon-Based MEMS Gyroscope Sense Closed Loop and Frequency Tuning Technology". The mode transfer function of dual-mass gyroscope detection is proposed, and the closed-loop controller with bandwidth expansion function is designed, furthermore, the frequency tuning method is introduced. In Chap. 6, "Temperature Influence on MEMS Gyroscope and Suppression Method", the temperature model, MEMS gyroscope temperature control and temperature compensation technology are introduced. Chapter 7 is "Dual-Mass Linear Vibration Silicon-Based MEMS Gyroscope Monitoring Circuit Design and Test Technology" which systematically introduces the test method of the MEMS gyroscope.

This book was written on the basis of the related research projects of the National Natural Science Foundation of China. I would like to express my gratitude to the institutions that have funded the contents of this book. The main content of this book was completed during the author's Doctoral stage in the Institute of Micro-inertial Systems and Devices, School of Instrument Science and Engineering, Southeast University and Georgia Institute of Technology, USA. In this process, I have received the help and support of Prof. Hongsheng Li, Dr. Paul Douglas Yoder, Prof. Shourong Wang, Prof. Libin Huang, Prof. Bailing Zhou, Prof. Bo Yang, Prof. Dunzhu Xia, Prof. Jinxing Liang and other comrades.. In the process of writing this book, Yong Yin, Cheng Luo, Xiaolei Wang, Jie Yan, Xiaolei Wang, Shuling Chen, Yunfang Ni, Lu Xu, Xi Zhu, Xia Sheng, Xukai Ding, Jia Liu and other comrades did a lot of useful work. I would like to express my heartfelt thanks to the above-mentioned comrades. This book refers to or directly quotes some domestic and foreign papers and works here to express gratitude to the authors of these papers and works. Finally, I would like to dedicate this book to my wife Dr. Zhongfang Zhang. Without her constant encouragement and care, it would be impossible for me to accumulate all the knowledge needed for writing this book.

This book involves multi-disciplinary fields. Due to the author's limited level, his research in some fields is not very in-depth. Therefore, it is inevitable that there are some inadequacies in this book.

Taiyuan, China Huiliang Cao

Contents

Chapter 1
Introduction

1.1 Silicon Based MEMS Gyroscope and Its Development

In the late 1980s, large scale and very large-scale integrated circuits appeared in the field of microelectronics science, making the integration of the circuit continuously improved, and the processing size is becoming smaller and smaller. On this basis, Micro Electromechanical System technology emerged after combining new technologies and methods. MEMS technology, as an emerging technology, integrates semiconductor manufacturing technology and micro-precision machining technology and other advanced manufacturing technology, covering many subject fields of microelectronics, materials, mechanics, chemistry and mechanics. Silicon micromechanical gyroscope is one of the important applications of MEMS technology in inertial navigation. It uses Coriolis effect to measure the input Angle rate of sensitive axis.

1.1.1 Silicon Based MEMS Gyroscope Advantages

As one of the core components of inertial navigation system, gyroscope has been developed for more than a century, including frame gyroscope, liquid float and air float gyroscope, flexible gyroscope, electrostatic gyroscope, laser gyroscope, fiber optic gyroscope and so on. As a new type of gyroscope, silicon micromechanical gyroscope has many advantages compared with traditional gyroscope [1, 2]:

(1) Low cost, mass production, short production cycle

The wafer level manufacture processes of silicon MEMS gyroscope structure and its integrated circuit are standardized, so that the MEMS gyroscope sensitive structure and the monitoring circuit can be mass produced and packaged in a small chip. At the same time, the standardized process flow can greatly increase the yield of silicon MEMS gyroscope (which is incomparable to traditional gyroscope), so as to reduce

H. Cao, *Dual-Mass Linear Vibration Silicon-Based MEMS Gyroscope*,
https://doi.org/10.1007/978-981-19-9247-6_1

the cost of a single gyroscope. At present, the price of a single silicon MEMS gyroscope in the market is about 1/1000 to 1/100,000th of the traditional gyroscope. At least three single-axis gyroscopes are often needed in the inertial navigation system, which enlarged the low cost advantage of silicon MEMS gyroscopes. In addition, because the silicon MEMS gyroscope has the characteristics of mass produced, once the model is formed, a large number of products can be processed in a short time, and can be quickly equipped, which is of great significance to national defense construction.

(2) Small size, light weight, low power consumption

The silicon MEMS gyroscope has the characteristics of small size, light weight and low power consumption because its sensitive structure is processed by MEMS technology and its monitoring circuit can also be realized by integrated circuit. Take the silicon MEMS gyroscope ADXRS150 produced by AD (Analog Device) Company as an example, its size is 7 mm × 7 mm × 3 mm, and its power consumption is about 40 mW [3]. Therefore, silicon MEMS gyroscope is more suitable for the occasion with strict requirements on volume, weight and power consumption. In practical applications, monolithic integrated three-axis micromechanical gyroscope can also be used to further improve its advantages in volume, weight and power consumption. In addition, redundant configuration schemes for multiple micro-gyros or gyro matrices can be used to improve the accuracy and reliability of the inertial navigation system.

(3) High reliability, long life, good impact resistance, good dynamic performance

Silicon MEMS gyroscopes have high reliability and long life capability, that is because they don't contain high-speed rotating rotor, so they can be regarded as a solid device without mechanical wear. In addition, their structure can be integrated with the electronic circuit, which greatly reduces the adverse effects of external interference. In addition, due to light weight and silicon material's good elasticity, the silicon MEMS gyroscope structures are with fast response, good dynamic performance and impact resistant ability. The silicon MEMS gyroscope even can bear 100,000 g's impact, this makes the silicon MEMS gyroscope can be used in harsh environment.

(4) Easy to digitize and intelligentize

The silicon MEMS gyroscope outputs analog signal, digital signal, frequency signal, etc., according to the special requirements of different inertial navigation systems. It can also be combined with MCU, with peripheral sensors and related algorithms, to achieve self-calibration, self-detection, self-compensation, and improve the environmental adaptability.

To sum up, because of the incomparable advantages over traditional gyroscopes, silicon MEMS gyroscope has broad applied prospects in military and civil fields, many countries have invested a lot of human, material and financial resources in its relative research work [4, 5]. In the military field, silicon MEMS gyroscope can not only be applied in micro-guided ammunition, intelligent ammo, ammunition safety insurance and detonating device, high G value penetration control of medium and short range tactical missiles, individual soldier combat system, UAV and other new fields. The high-precision silicon MEMS gyroscope can also replace traditional gyroscopes in the stabilization of tactical search heads, autopilot, short and medium range missiles, torpedo fuze, low-cost attitude heading reference systems, and small, air-dropped or gun-fired, midcourse and terminal precision guided submunitions. In the civil field, silicon MEMS gyroscopes can be applied in consumer electronics, automotive engineering, mobile communications, geodesy, geological exploration, robots, vibration monitoring, inertial mouse, stability control of cameras, toys, sports equipment, in body monitoring, wheelchairs and other fields [6–11].

1.1.2 Silicon-Based MEMS Gyroscope Classification

With the development of research, the shape and form of silicon MEMS gyroscope sensitive structure tends to diversify, but no matter how the change, the Coriolis effect was utilized as the working principle. The Coriolis effect is generated when a vibrating mass block is rotated by an angular velocity, and the Coriolis force is proportional to the angular velocity. According to different working modes, silicon MEMS gyroscopes can be divided into vibration type, rotor type and medium type [1]. Vibration type silicon MEMS gyroscope is the mainstream of development, which can be divided into linear vibration type and angular vibration type according to vibration mode. According to the driving method can be divided into electrostatic drive, electromagnetic drive, the detection method can be divided into capacitance detection, current detection, frequency detection, resistance detection; According to the quantity of mass block, it can be divided into single-mass block, dual-mass block and multi-mass block. On the basis of the above classification, the dual-mass linear vibration silicon-based MEMS gyroscope structure has become the mainstream silicon micromachined gyroscope structure due to its good resistance to detecting axial acceleration disturbance and easy processing (the structure only moves in plane). Therefore, this book mainly studies the dual-mass linear vibration silicon-based MEMS gyroscope . The silicon micro-gyroscope adopts capacitive drive and capacitive detection mode, which mainly includes two parts: micro-sensitive structure and measurement and control circuit. Among them, the micro-sensitive structure is made of silicon material by micro-machining, which mainly includes driving and detecting modes. The measurement and control circuit provides necessary working state and extracts working signal for the gyro, which is mainly composed of driving loop and detecting loop. In the working process of the silicon micromechanical gyroscope, the driving loop ensures constant amplitude vibration of the driving mode, and the angular velocity around the sensitive axis causes the Coriolis mass to produce Coriolis force in the direction of the detection axis (perpendicular to the driving and sensitive axes).Under the influence of Coriolis force, the detection frame will generate a displacement related to the input Angle rate as the input of the detection circuit, and the detection circuit will go through a series of processing and finally output.

1.2 Silicon Based MEMS Gyroscope Development

Since the first silicon micromechanical gyroscope was developed by Draper Laboratory in 1988, the precision of silicon micromechanical gyroscope has been greatly improved with the continuous optimization and improvement of new structure and measurement and control methods. In recent years, foreign institutions engaged in silicon micromechanical gyroscope research mainly include Seoul National University in South Korea [12–16]. University of California, Irvine [17–25] and Berkeley [26–28], Michigan State University [29], Jet Power Laboratory [30–33], Carnegie Mellon University [34–37], Georgia Institute of Technology [38–41], Draper Laboratory [42–45], Honeywell [46–49], LITEF company [50], Bosch Company [51, 52], HSC-IMIT Institute [53], University of British Columbia, Canada [54–56], Delft University, Netherlands [57], Middle East University of Science and Technology, Turkey [58–60], University of Helsinki, Finland [61–63], UK [64],THALES In France [65], Italy [66], Thailand [67], Pakistan [68], etc. Domestic research in this area is also being carried out, mainly including: Southeast University [69–74], Tsinghua University [75], Peking University [76–80], North University of China [81, 82], Harbin Institute of Technology [83, 84], National University of Defense Technology [85–88], Northwestern Polytechnical University [89], Nanjing University of Science and Technology [90–92], Beijing University of Aeronautics and Astronautics, Zhejiang University, Shanghai Institute of Microsystems, 55 and 13 of the China National Telecommunication Industry, 13 and 33 of the Aerospace industry, and 618 of the aviation industry.

Because the domestic research in this field started late compared with the developed countries, coupled with the gap between domestic processing technology and foreign countries, even there are some gaps with the advanced level of foreign countries. However, with the continuous investment in micromechanical system industry during the ninth, fifteenth and eleventh Five-Year plan period, China has made great progress in theoretical research and engineering prototype in the field of micromechanical system, gradually narrowing the gap with foreign countries.

After a detailed summary of the literature published by various research institutions at home and abroad, the book holds that the development of linear vibrating silicon micromechanical gyroscopes can be roughly divided into three stages [20, 58, 59, 62, 93, 94]. The first stage is the design stage of basic structure and measurement and control circuit. The main structure and the principle of measurement and control circuit verification and improvement. In this stage, driving closed loop and detecting open loop are usually adopted. However, due to the limitation of machining technology, there are various non-fundamental errors (especially quadrature errors) in detecting modal signals, which restrict the accuracy of gyro to a great extent. The second stage is quadrature error elimination and force feedback control. Under the current processing conditions, the quadrature error and force feedback can be eliminated effectively by adding nulling and feedback mechanism in the basic structure and matching with the relevant controller. In this stage, the static and dynamic performance of gyro will be greatly improved. The third stage is the matching of gyro drive and detecting mode resonant frequency. In order to obtain a better SNR, it is necessary to improve the mechanical sensitivity of the structure. Usually, the difference between the resonant frequencies of the driving and detecting modes is inversely proportional to the sensitivity, so the maximum mechanical sensitivity can be obtained when the two modes are matched.

1.2.1 Overview of Silicon Based MEMS Gyroscope Abroad

The MEMS gyroscope structure proposed by Seoul National University in 2007 is shown in Fig. 1.1 [12–15], and its structure is in the form of single-mass fully decoupled linear vibration. From the distribution of electrodes, it can be seen that the gyro adopts push-pull drive and differential detection mode, and the detection feedback electrode in the structure can realize the closed-loop control of the detection loop. This kind of structure can not only obtain larger driving force, but also reduce the common mode interference and increase the SNR in the detection channel. In structural processing, deep reactive ion etching (DRIE) technology is adopted, and vacuum packaging improves the quality factor. Phase locked loop (PLL) controller is adopted in the driving loop, and PLL and automatic gain controller (AGC) are introduced into the detection loop to realize the closed-loop control of the detection loop [13]. In the frequency tuning part, the detection mode of gyro is used as voltage-controlled oscillator module in PLL, and the input signal of phase discriminator is composed of driving excitation signal and detecting displacement signal. Due to the small difference between the resonant frequencies of the driving and detection modes, reasonable configuration of tuning controller parameters can match the frequencies of the two modes [15].

Bosch reported a dual-mass linear vibration silicon-based MEMS gyroscope DRS-MM3 [51, 52] for automotive systems. Its structure adopts tuning fork vibration form, and the left and right mass blocks are coupled by an intermediate connecting beam. In the measurement and control system, using the form of digital integrated circuit, digital processing part is mainly composed of driving closed-loop loop, detection closed-loop (detection closed-loop control mode), output compensation loop. Among them, the driving closed-loop loop adopts the driving mode of phase-locked loop and automatic gain control loop, and the output compensation loop adopts the combination of temperature sensor and regulator to sense the internal temperature of gyro and compensate the output signal of gyro detection loop. The compensated signal is output through SPI bus, as shown in Fig. 1.2.

A four-mass linear vibration silicon-based MEMS gyroscope proposed by the University of California, Irvine, is shown below. In order to reduce the mechanical thermal noise and improve the signal-to-noise ratio of the output signal of the structure, a high vacuum degree is adopted (the mechanical thermal noise of the substrate is about 10°/h RMS in air, but it increases to 0.01°/h RMS in vacuum). In terms of measurement and control system, the gyroscope adopts driving loop and detection

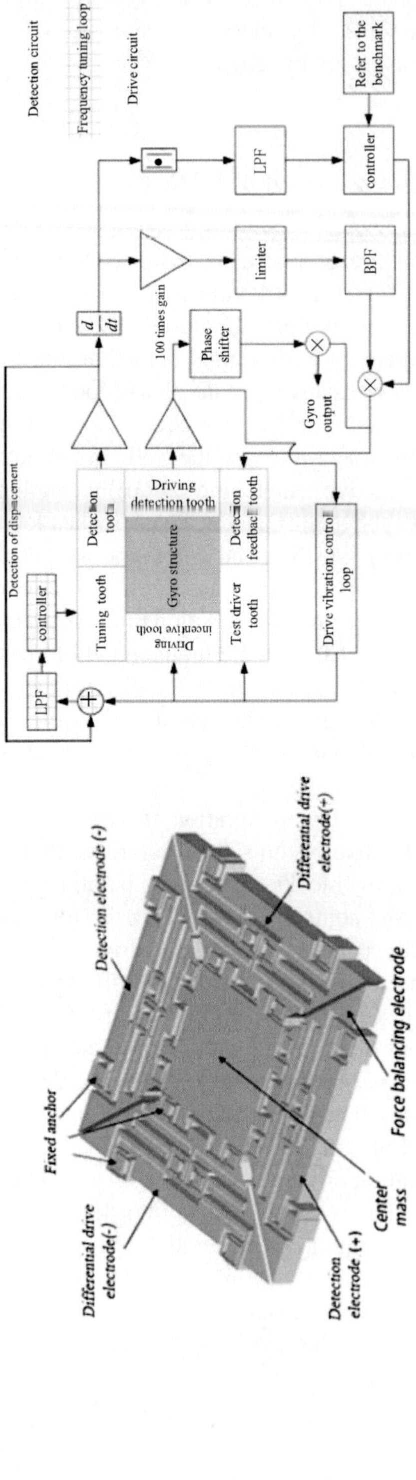

Fig. 1.1 Micromechanical gyro structure and measurement and control system, Seoul National University

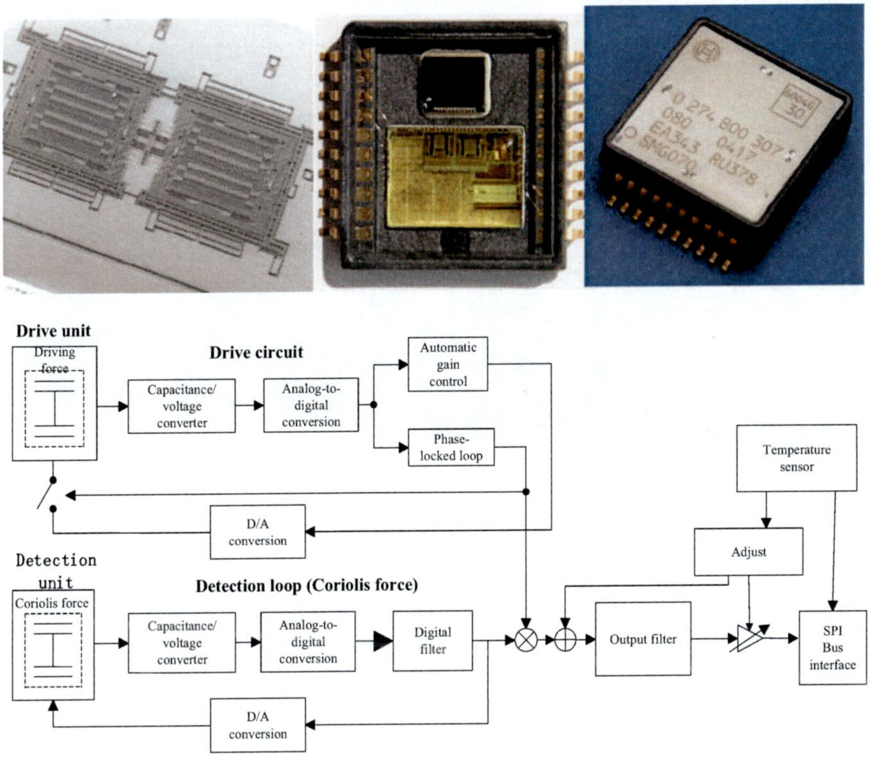

Fig. 1.2 Micromechanical gyro structure, chip and measurement and control system proposed by Germany Bosch company

loop based on DSP platform, in which the driving mode adopts phase-locked loop and automatic gain control, to ensure the stable work of the system (Fig. 1.3).

The THALES come up with a dual-mass linear vibration silicon-based MEMS gyroscope [65, 95] in 2009, as shown in Fig. 1.4. The whole structure is machined by DRIE process on SOI substrate. The structure contains drive excitation, drive detection, detection force feedback, detection, quadrature nulling and frequency tuning comb. The machined structure is encapsulated in HTCC ceramic shell with high vacuum degree. The quadrature offset signals of 450°/s can be cancelled by the quadrature nulling combs distributed at four corners under the external voltage of 15 V. In addition, on frequency-tuned combs, the electrostatic stiffness generated by a voltage of 15 V eliminates the modal frequency difference of 500 Hz. Figure 1.4 shows the overall schematic diagram of micromechanical gyro digital measurement and control system of THALES Company in France. Compared with analog circuit, it has the advantages of low noise, flexible controller parameter setting and more complex control algorithm. As can be seen from the figure, the gyro driving loop is realized based on digital phase-locked loop, and the nulling of quadrature error consists of two parts: the quadrature compensation force applied on the comb teeth by the detection force feedback and the electrostatic negative stiffness on the comb

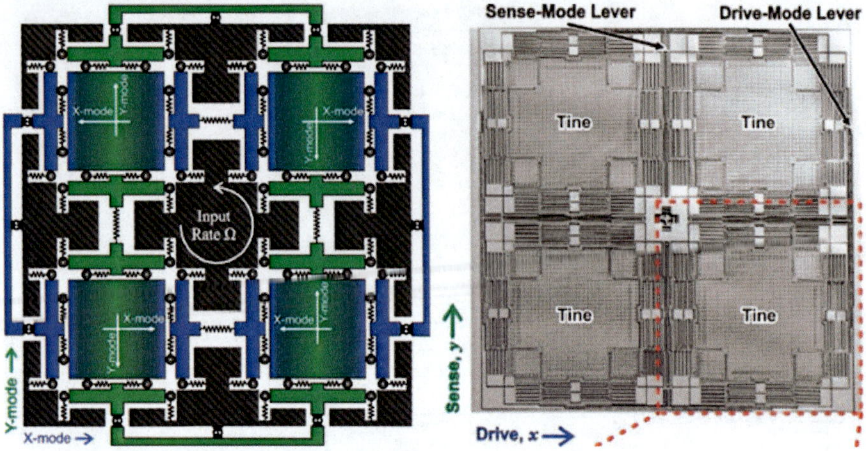

Fig. 1.3 Micromechanical gyro structure and photo presented by University of California, Irvine, USA

teeth by the quadrature nulling. The system adopts the detection closed-loop control loop based on PI controller, and the frequency tuning module adopts the open-loop working principle and is controlled by a fixed voltage generator. At the same time, the temperature compensation of the gyroscope is carried out in the third order.

LITEF GmbH, Germany, has proposed a dual-mass gyro structure for high precision applications. Its schematic diagram is shown in Fig. 1.5 [50, 96, 97]. The driving mode of the measurement and control system of the gyro adopts PLL controller, and the detection mode is closed-loop mode. At the same time, the quadrature nulling loop based on quadrature force nulling method and quadrature stiffness compensation method is also included. In modal matches, the gyro signal source in the quadrature signal channel by the aid of external input to the gyro testing feedback tooth, the signal is detected by gyroscope and then extracted by Coriolis signal channel, produce control signals after being auxiliary signal in phase demodulation match two modal resonance frequency channel of the auxiliary signal detection (modal matching maximum peak). The third order polynomial temperature compensation method is adopted in the system.

In 2010, Sensonor AS of Norway reported on its butterfly high-precision gyroscope [98]. Its structure and electrode distribution are shown in Fig. 1.6, and its structure is equipped with quadrature nulling electrodes. The whole system adopts digital circuit design, which is mainly composed of two closed-loop loops of drive and force feedback. Adopted the differential charge amplifier with low noise detection of target capacitance, The detection signals are then converted into bitstreams by a fifth order Σ-Δ analog-to-digital converter into a digital processor, by correlation algorithm to get the corresponding control signal, signal after the bitstream buffer and filter applied to the gyro structure, related testing data are shown in Table 1.1. The company's STIM202 micromechanical gyroscope was awarded the European

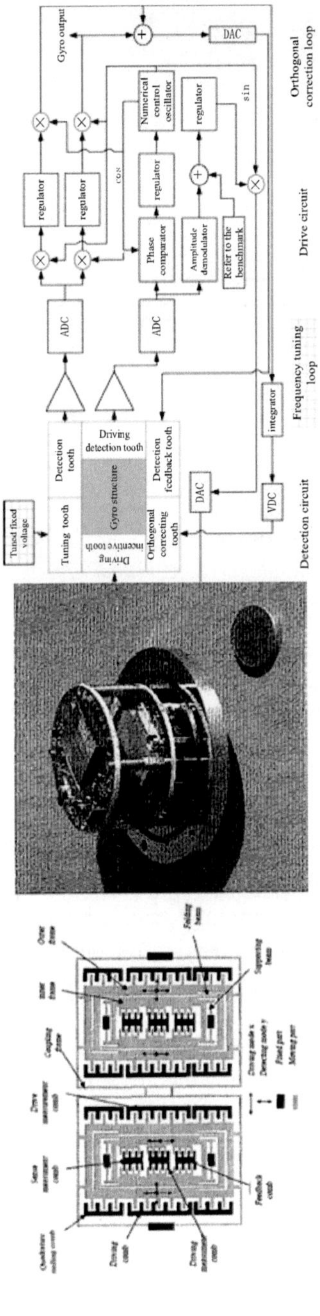

Fig. 1.4 THALES micromechanical gyro structure, prototype and measurement and control system

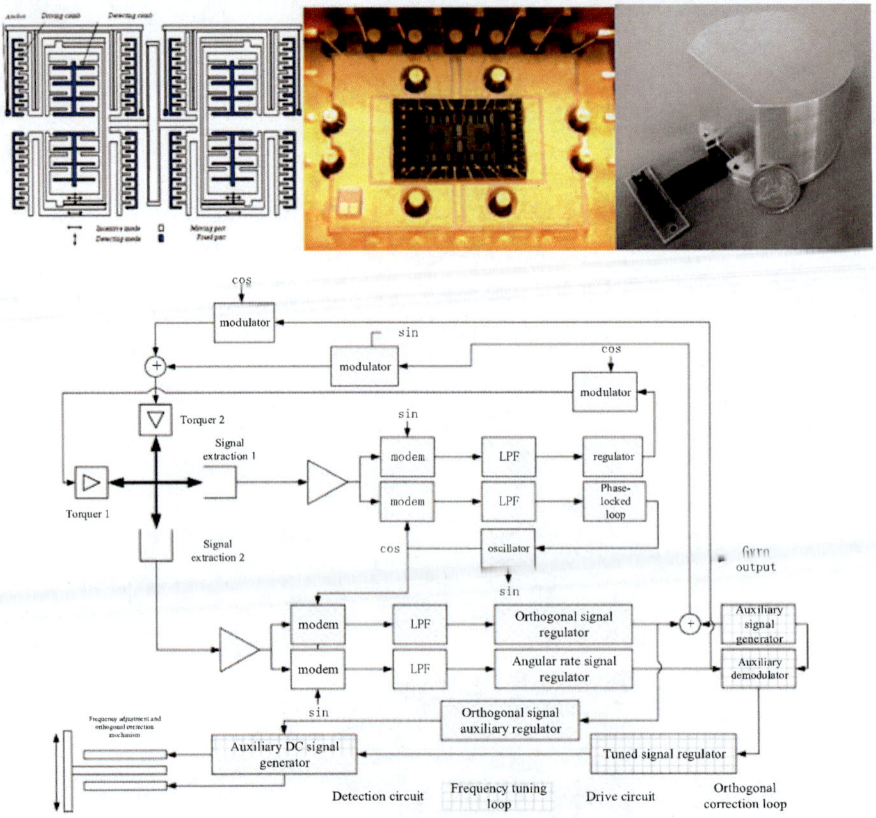

Fig. 1.5 LITEF micromechanical gyro structure and product photos

Invention Award for High Precision MEMS Gyroscope in 2010 for its high accuracy (zero offset stability up to 0.5°/h) and high stability.

In 2009, Georgia Institute of Technology realized automatic mode matching on the basis of undecoupled dual-mass linear vibration silicon-based MEMS gyroscope structure [40], and the schematic diagrams of its structure and measurement and control circuit are shown in Fig. 1.7 respectively. The gyro structure is equipped with an quadrature nulling electrode to suppress the influence of quadrature errors, and the electrode is used as a frequency tuning electrode (the electrode is in the pressure film mode in the detection mode direction). Digital PLL method was used to control the drive loop, reference direct flow superposition control quantity was used for quadrature calibration voltage, and open-loop working mode was used for detection mode. According to the literature in terms of frequency tuning in 0 Hz frequency offset mechanical sensitivity, noise level and zero bias stability than modal mismatch condition improved, and through test of quadrature signal size modal matching state (modal matching quadrature signal peak), but the larger the quadrature error will affect the precise matching mode. The detection frequency of 100 Hz can be changed when the tuning voltage is 45 V, and the quadrature nulling voltage is within ±20 V.

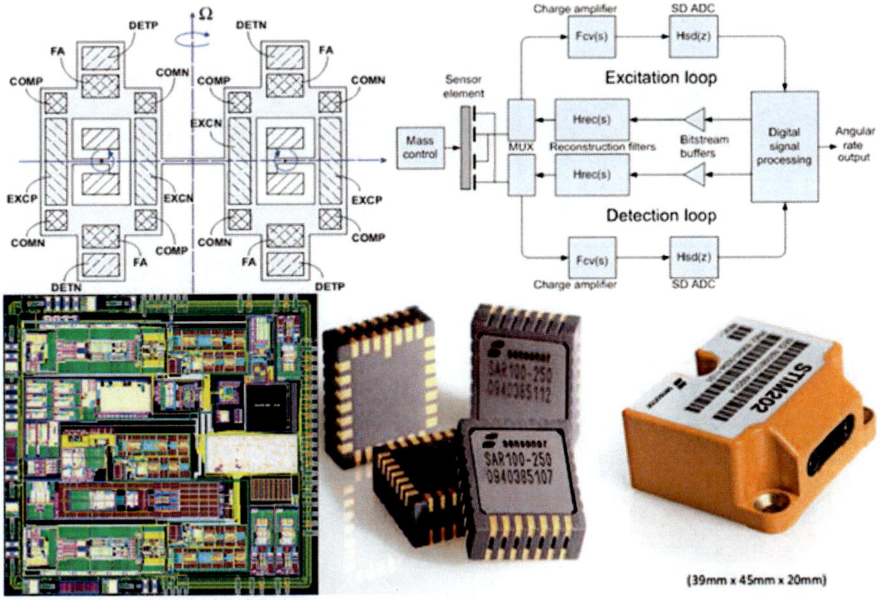

Fig. 1.6 The MEMS gyroscope structure and monitoring system of Sensonor AS

Middle East University of Science and Technology in Turkey reported a single-mass fully decoupled silicon microgyroscope in 2008, and developed a driving closed-loop loop based on AGC self-excited oscillation technology based on this structure [99]. In 2012, the project team proposed a new dual-mass line vibration decoupled structure, and the schematic diagram of the structure of a single mass block is shown in Fig. 1.8 [58]. The structure contains an quadrature nulling comb that can correct most of the quadrature coupling stiffness, and the remaining quadrature signals are used in the modal matching loop: quadrature signal is quadrature signal demodulation with the drive signal phase (driving signal and the modem, low-pass filter can be seen as phase discriminator of phase-locked loop), get the phase difference between the two signal information, the information in the form of voltage input to the PI controller, the controller output frequency tuning of dc control signal is sent to the gyro public mass. In addition, it is pointed out that if the quadrature signal is not weakened below a certain limit, the frequency tuning controller will always be saturated and cannot achieve the purpose of frequency tuning [60]. The frequency tuning scheme in this design is relatively efficient, and can be used to change the detection mode resonant frequency of 1000 Hz with ±2.5 V DC voltage. The PI controller in the detection closed loop can effectively tune the gyroscope bandwidth limited by the mechanical frequency difference. In this paper, the mechanical bandwidth is only 4 Hz, which is later adjusted to 50 Hz by the controller.

Table 1.1 Performance summary of gyro in some domestic and foreign research institutes

Parameter names	Research Units											
	SEOUL NATL UNIV	Bosch	Irvine	THALES	LITEF	SensoNor	Georgia Institute of Technology	Middle East Technical University	Peking University	Nanjing institute of technology	Southeast University	National University of Defense Technology
Measurement and control circuit platform	Simulation PCB	Digital ASIC	digit circuit	digit circuit	digit circuit	Digital ASIC	Digital ASIC	Simulation PCB	Simulation PCB	Simulation PCB	FPGA	Simulation PCB
Operation mode of drive loop	PLL	Digital PLL	Digital PLL	Digital PLL	Digital PLL	Digital AGC	Digital PLL	AGC	AGC	AGC	Digital PLL	PLL
Quadrature nulling way	Force nulling	–	–	Force and stiffness nulling	Force and stiffness nulling	Force and stiffness nulling	Stiffness nulling	Stiffness nulling	Force nulling	–	Stiffness nulling	Stiffness nulling
Closed loop detection mode	PID controller	Digital	Digital	Digital	Digital	Σ-Δ	Open loop	Simulation PI	Simulation PI	–	Digital PID	Simulation PI
Tuning principle	Closed loop	–	Closed loop	Open loop	Closed loop signal assist	–	Closed loop quadrature assist	Closed loop PLL control	Open loop tuning	–	–	–
Temperature compensation	–	Digital	-	The third order number	The third order number	–	–	–	–	Simulation	Digital	–
Quality factor	357	–	1170000	50000	1000	–	30000	48586	570	100000	1000	8000

(continued)

Table 1.1 (continued)

Parameter names	Research Units											
	SEOUL NATL UNIV	Bosch	Irvine	THALES	LITEF	SensoNor	Georgia Institute of Technology	Middle East Technical University	Peking University	Nanjing institute of technology	Southeast University	National University of Defense Technology
resonant frequency(Hz)	7816	15000	2000	10500	10000	10000	15000	14000	9000	20000	3700	2000
structure height(μm)	80	–	50	60	50	–	60	100				
range (°/s)	>500	–	–	1000	1000	500	150	–	200	500	300	200
Scale factor (mV/°/s)	3.8	–	–	10^6 LSB/°/s	–	–	88	–	41	21.5	1	11.7
bandwidth(Hz)	~70	60	–	300	500	–	1-10	50	94	100	–	50
zero-bias stability(°/h)	–	1.35	0.88	<0.1	0.12	0.04	0.15	0.83	4	7.7 (1σ)	8 (1σ)	23.4 (1σ)
Angle random walk(°/$h^{0.5}$)	0.084	0.147	0.06	0.01	0.3	0.002	0.003	0.026	0.17	–	0.06	–
shock resistance(g)	–	–	–	20000	4000	–	–	–	–	–	–	–

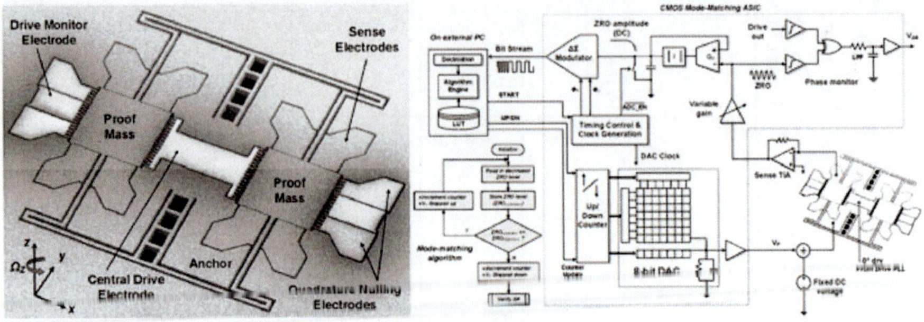

Fig. 1.7 The MEMS gyroscope structure and monitoring system of Georgia Institute of Technology

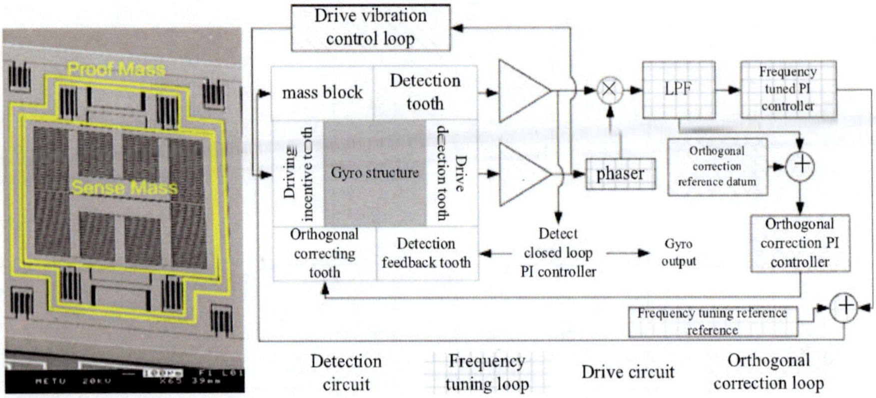

Fig. 1.8 Middle East University of Science and Technology micromechanical gyro structure and measurement and control system

1.2.2 Overview of Silicon Based MEMS Gyroscope in China

In 2010, Peking University proposed a single-mass linear vibration silicon-based MEMS gyroscope structure, as shown in Fig. 1.9 [78]. The structure first worked in the open-loop state [100], and then designed a detection closed-loop control loop based on it, using the sixth-order Σ–Δ, to obtain a signal-to-noise ratio of 100dB in the full range. The noise of output signal is $-90\text{dBV/Hz}^{0.5}$. In 2011, the research group designed an quadrature nulling and detection closed-loop loop based on the detection force feedback comb on its single-mass structure, and expanded the bandwidth of the gyro by combining notch and PID phase advance controller. The effective bandwidth of the gyro increased from 30 Hz of the open-loop to 98 Hz of the closed-loop. At the same time, the scale factor of gyro decreased from 27.1 mV/°/s of open-loop to 7.1 mV/°/s [77]. In addition, a simplified closed-loop PID advance nulling controller for its single-mass gyro was proposed [101, 102].

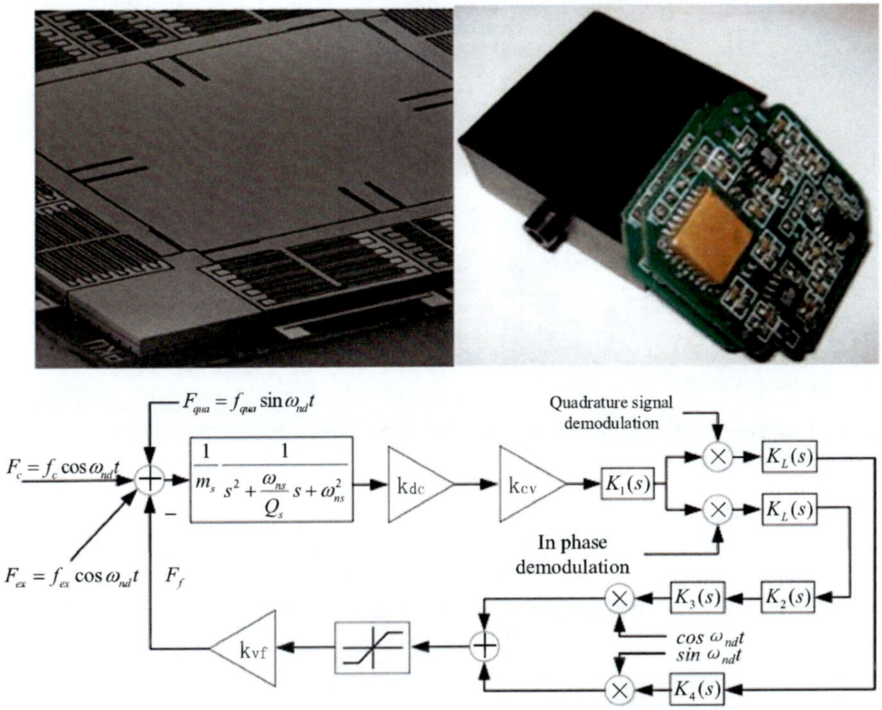

Fig. 1.9 Photo and measurement and control system of gyro structure proposed by Peking University

In 2013, Nanjing University of Science and Technology proposed a dual-mass linear vibration, decoupled silicon-based MEMS gyroscope structure and its measurement and control circuit (see Fig. 1.10) [103]. The driving loop adopts self-excited oscillation AGC closed-loop, and the detection loop is open loop. The meter head package adopts LCC12 surface mount ceramic tube shell, the solder is eutectic gold antimony alloy, the volume is 15 × 15 × 3.5 mm^3, the package is vacuum, the overall volume of the gyroscope is 31 × 31 × 12 mm^3, power consumption is 288 mW, the scale factor is nonlinear, symmetry and repeatability are 37 ppm, respectively. 184 ppm and 155 ppm, zero bias repeatability is 12°/h, threshold and resolution are 0.008°/s.

In 2014, Southeast University proposed a dual-mass linear vibration silicon-based MEMS gyroscope structure with quadrature nulling comb teeth and detection feedback teeth, as shown in Fig. 1.11. Quadrature nulling and detection closed-loop control loops were designed respectively based on FPGA and analog circuit [104]. Among them, FPGA mode adopts the driving loop adopts PLL and AGC collocation control method to realize the frequency locking and amplitude stabilization of the driving mode, and PI controller is used to realize the closed-loop control of the detection loop. Coupling stiffness nulling is adopted in quadrature nulling mode, digital PID phase advance nulling is adopted in feedback controller, and temperature

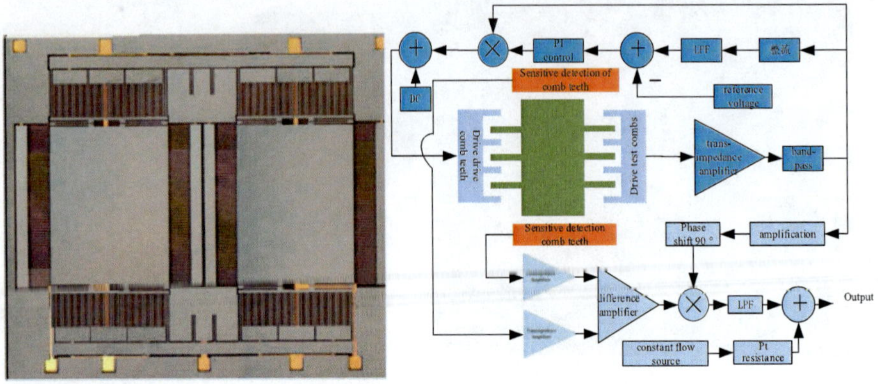

Fig. 1.10 Structure and measurement and control system of double mass gyro in Nanjing University of Science and Technology

compensation is carried out for zero deviation and scale factor of gyroscope. The simulation scheme uses its own oscillation mode to build the driving loop, and the phase leading proportional integral controller is used in the detection loop to expand the bandwidth.

In 2011, the butterfly gyro structure developed by National University of Defense Technology is shown in Fig. 1.12 [85]. In 2013, the quadrature nulling electrode [88] was added on the basis of this structure, and the stiffness nulling method was adopted to suppress the quadrature error. After the nulling, the performance of gyro was

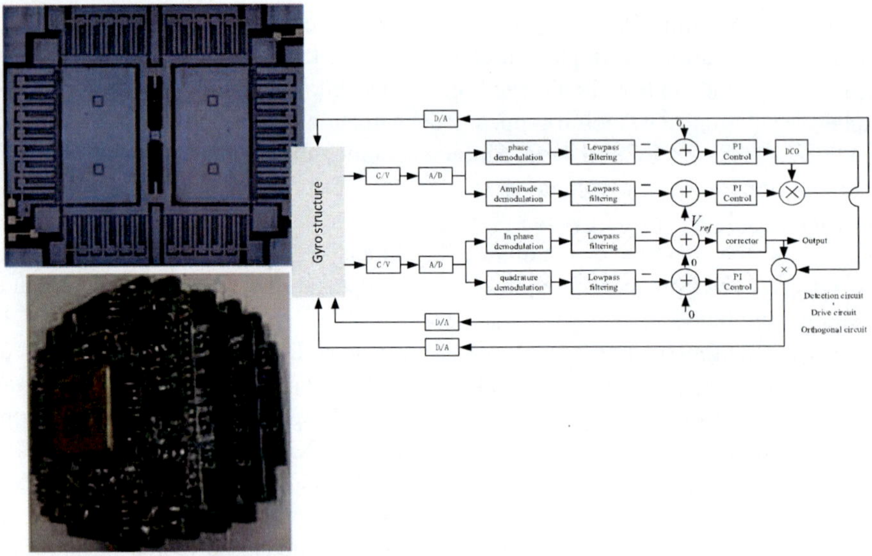

Fig. 1.11 Southeast University double mass gyro structure and analog circuit-based measurement and control system

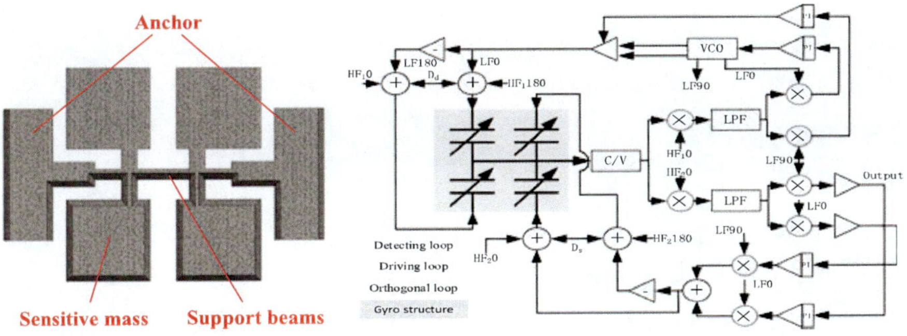

Fig. 1.12 Butterfly gyro structure and measurement and control system of National University of Defense Technology

significantly improved, and the drift trend was improved in the open-loop detection state, and the zero-skewness stability was increased from 89°/h to 17°/h. The scale factor temperature stability decreased from 622 to 231 ppm/°C. In addition, the research group realized detection closed-loop control based on PID controller in the later work. At the same time, the quadrature force nulling method was adopted to eliminate the quadrature error. The zero-skew stability of the system decreased from 136 to 23°/h, and the nonlinear degree of scale factor decreased from 0.67 to 0.2%. The temperature stability of the scale factor was optimized from 664 to 382 ppm/°C. After the closed-loop detection, the vibration test of the system was carried out, and the anti-vibration characteristics of the system were significantly improved [87]: in the same vibration process, the output peak value of the open-loop gyroscope was 3.1 V while that of the closed-loop was 8 mV, reflecting the good dynamic performance of the closed-loop state.

Tsinghua University has also carried out research on micromechanical gyroscope. The picture of the prototype is shown below. The measurement and control system are built on the FPGA hardware platform, and the temperature coefficient of the scale factor is effectively improved through the automatic compensation algorithm of the scale factor (Fig. 1.13).

1.3 Silicon Based MEMS Gyroscope Design Process

Book that the design process of the silicon micromechanical gyroscope is a constantly found the problem, the process of continuous improvement, this process is mainly limited to processing technology, the application environment, encapsulation and a number of factors such as measurement and control technology, and from the whole need weighed from three directions: structure direction, encapsulation direction and the direction of the circuit, schematic diagram as shown in Fig. 1.14.

Can be seen from the above, around the application environment and the core indicators demand, first determine the structure design, packaging (cost), electricity

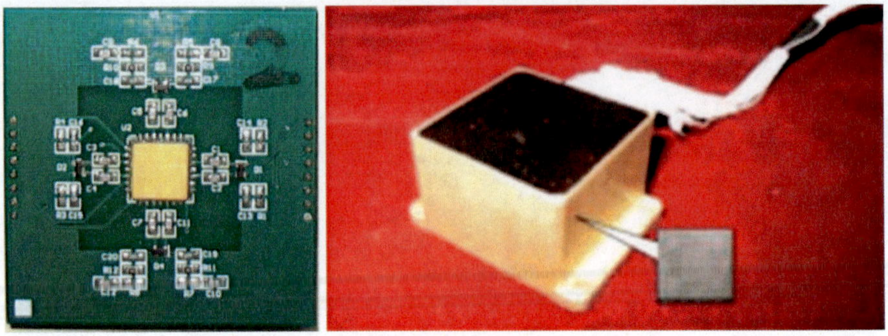

Fig. 1.13 Micromechanical gyroscope prototype of Tsinghua University

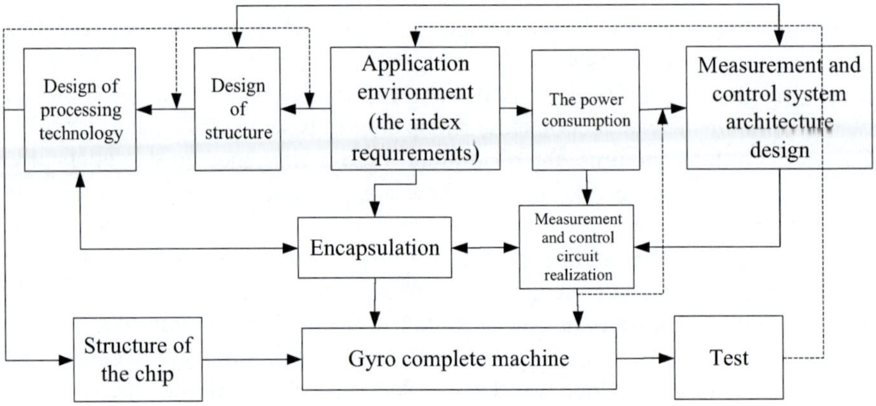

Fig. 1.14 Schematic diagram of micromechanical gyroscope design process

power consumption of the three directions, such as the applications of gun ammunition environment, need structure with high overload resistance properties, need encapsulation has the quality of small, reliable connection, the characteristics of easy to mass production, The power supply should be missile-borne and requires less power consumption to save missile-borne power resources. The dotted lines in the figure above represent feedback, which means the design needs to be adjusted accordingly. After the completion of the structural design, the production and processing shall be carried out according to the existing or cost appropriate process, and the processing process shall also be limited by the form of packaging. The satisfactory processing process requires long-term and multi-round debugging, during which the structural design parameters may be adjusted according to the maturity of the process. Better or more mature processing technology should ensure a higher yield to reduce processing costs. Gyroscope measurement and control system architecture design work should be synchronized structure design process, measurement and control system including the structure of signal extraction methods, the size of the drive, drive circuit, whether to need to increase the quadrature nulling system, whether to need to adjust the modal frequency, if the need for testing circuits closed-loop, mechanical sensitivity, whether need to expand the bandwidth, and so on questions. Based on measurement and control system architecture is built, the combination of encapsulation and work out the structure of the chip of the gyroscope (sample) machine, further according to the requirement of the application environment of gyro machine test, test results directly reflect the gyroscope prototype can meet the demand, if can't meet the needs of the demand, it's need to improve the individual link and redesign.

Due to the structure design process and machining process are relatively mature, and There are many excellent books on these topics, the measurement and control system architecture design is relatively complex, so the part as an example to conduct a preliminary analysis, sorting out design ideas, and thereby part design as the main content in detail in the following sections. In general, the static characteristics of high-precision silicon micromechanical gyroscope should include small zero value and drift, high mechanical sensitivity and signal-to-noise ratio, and at the same time, it should have a wide working bandwidth and small detection displacement and other excellent dynamic characteristics. Figure 1.15 shows the influence of gyro structure on its static and dynamic characteristics.

From the figure above, it can be seen that there are unfavorable factors in the structure, such as the primary factor and the machining error, among which the former includes the high sensitivity and the wide bandwidth in the modal frequency difference factor are contradictory, usually in the design of the structure to ensure the

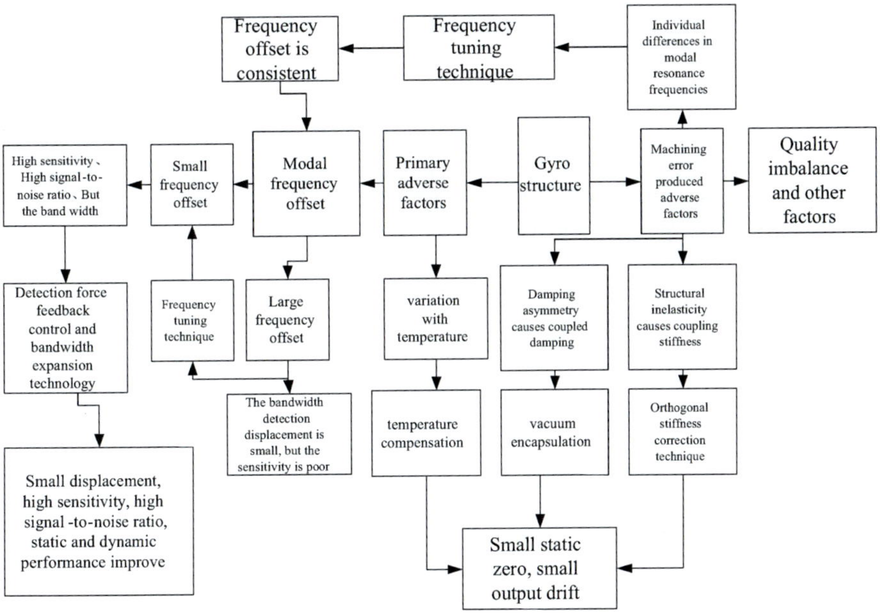

Fig. 1.15 Schematic diagram of influence of structure on static and dynamic characteristics of silicon micromechanical gyroscope

bandwidth at the expense of high sensitivity, which is also one of the main bottlenecks limiting the accuracy of gyro. In addition, many literatures point out the great influence of processing error of gyro meter performance, the machining error leaded to unfavorable factors mainly includes coupling stiffness due to structural elastic inequality, gyroscopic modal frequency offset between the individual differences, damping asymmetry leaded to the coupling damping and uneven quality, the influence of the former is more apparent. The above problems can be solved by improving the processing technology, but this needs a longer cycle and higher cost, which is in contradiction with the urgent needs of domestic (especially military industry). Therefore, on the basis of the existing technology, it is necessary to improve the precision of gyro by designing a reasonable measurement and control system architecture.

Chapter 2
Silicon Based MEMS Gyroscope Structure and Working Principle

2.1 Chapter Introduction

Dual-mass linear vibration silicon-based MEMS gyroscope is an inertial sensor based on the principle of Coriolis effect. Therefore, the principle of its operation should be studied in depth to find out the main factors and errors and interference items in the principle level. On the basis of the above, relevant vibration models and motion modes are extracted to optimize the subsequent control system. This chapter mainly studies the basic theory of silicon micromechanical gyroscope, and introduces the design and machining technology of gyroscope structure.

2.2 Silicon Based MEMS Gyroscope Mechanics Principle

2.2.1 Coriolis Acceleration

Coriolis acceleration, also known as Coriolis acceleration, was first proposed by G. G. Coriolis in 1835. It describes the interaction between the motion of a particle in a moving coordinate system and the rotational motion of a moving coordinate system in an inertial coordinate system, its magnitude is proportional to the velocity of the particle in the moving coordinate system and the angular velocity of the moving coordinate system with respect to the inertial coordinate system, and the direction is perpendicular to the plane composed of both.

In Fig. 2.1, the position of the inertial coordinate system $n\text{-}x_n y_n z_n$ $n - x_n y_n z_n$ (hereinafter referred to as n system) is fixed, and the shell coordinate system $b\text{-}x_b y_b z_b$ (hereinafter referred to as B system) has an implicated motion relative to the inertial coordinate system. the implicated motion is a rotary motion of angular velocity is ω_{nb}. The position vector of point d with respect to the inertial and shell coordinates are $\overrightarrow{r_{nd}}$ and $\overrightarrow{r_{bd}}$. In the N system:

H. Cao, *Dual-Mass Linear Vibration Silicon-Based MEMS Gyroscope*,
https://doi.org/10.1007/978-981-19-9247-6_2

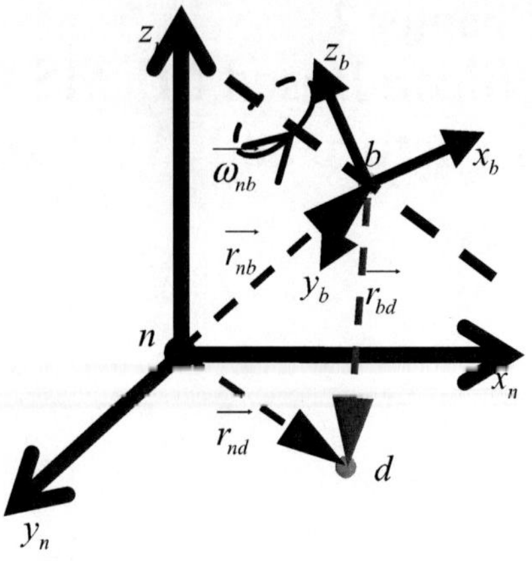

Fig. 2.1 Schematic diagram of Coriolis acceleration generation

$$\overrightarrow{r_{nd}} = r_{ndx}\overrightarrow{i_n} + r_{ndy}\overrightarrow{j_n} + r_{ndz}\overrightarrow{k_n} \tag{2.1}$$

the $\overrightarrow{i_n}$, $\overrightarrow{j_n}$ and $\overrightarrow{k_n}$; r_{ndx}, r_{ndy} and r_{ndz} are the unit vectors in the x_n, y_n, and z_n directions of the n-series and the coordinates of d in those three directions. Take the first and second derivatives of the left and right sides of the equation with respect to time and we can get the velocity $\overrightarrow{v_{nd}}$ and acceleration $\overrightarrow{a_{nd}}$ of point d in N system.

$$\overrightarrow{v_{nd}} = \frac{\mathrm{d}_n r_{ndx}}{\mathrm{dt}}\overrightarrow{i_n} + \frac{\mathrm{d}_n r_{ndy}}{\mathrm{dt}}\overrightarrow{j_n} + \frac{\mathrm{d}_n r_{ndz}}{\mathrm{dt}}\overrightarrow{k_n} \tag{2.2}$$

$$\overrightarrow{a_{nd}} = \frac{\mathrm{d}_n^2 r_{ndx}}{\mathrm{dt}^2}\overrightarrow{i_n} + \frac{\mathrm{d}_n^2 r_{ndy}}{\mathrm{dt}^2}\overrightarrow{j_n} + \frac{\mathrm{d}_n^2 r_{ndz}}{\mathrm{dt}^2}\overrightarrow{k_n} \tag{2.3}$$

In the same way, we can get the expression of location $\overrightarrow{r_{bd}}$, velocity $\overrightarrow{v_{bd}}$ and acceleration $\overrightarrow{a_{bd}}$ of point d in the B system.

$$\overrightarrow{r_{bd}} = r_{bdx}\overrightarrow{i_b} + r_{bdy}\overrightarrow{j_b} + r_{bdz}\overrightarrow{k_b} \tag{2.4}$$

$$\overrightarrow{v_{bd}} = \frac{\mathrm{d}_b r_{bdx}}{\mathrm{dt}}\overrightarrow{i_b} + \frac{\mathrm{d}_b r_{bdy}}{\mathrm{dt}}\overrightarrow{j_b} + \frac{\mathrm{d}_b r_{bdz}}{\mathrm{dt}}\overrightarrow{k_b}. \tag{2.5}$$

$$\overrightarrow{a_{bd}} = \frac{\mathrm{d}_b^2 r_{bdx}}{\mathrm{dt}^2}\overrightarrow{i_b} + \frac{\mathrm{d}_b^2 r_{bdy}}{\mathrm{dt}^2}\overrightarrow{j_b} + \frac{\mathrm{d}_b^2 r_{bdz}}{\mathrm{dt}^2}\overrightarrow{k_b} \tag{2.6}$$

$\overrightarrow{i_b}$, $\overrightarrow{j_b}$ and $\overrightarrow{k_b}$; r_{bdx}, r_{bdy} and r_{bdz} are the unit vectors in the x_b, y_b and z_b directions of the B system and the coordinates of point d in these three directions respectively.

When $\overrightarrow{\omega_{nb}} = \overrightarrow{0}$, there is no relative motion between the two frames, the motion of point d with respect to points n and b is the same, then $\overrightarrow{v_{nd}} = \overrightarrow{v_{bd}}$, $\overrightarrow{a_{nd}} = \overrightarrow{a_{bd}}$. When $\overrightarrow{\omega_{nb}} \neq \overrightarrow{0}$, The implicated motion of B series relative to N series is rotational motion. Let's say the vector from n to B is $\overrightarrow{r_{nb}}$, then:

$$\overrightarrow{r_{nd}} = \overrightarrow{r_{nb}} + \overrightarrow{r_{bd}} \tag{2.7}$$

Take the derivative of this in the N series, then we can get:

$$\frac{\mathrm{d}_n \overrightarrow{r_{nd}}}{\mathrm{dt}} = \frac{\mathrm{d}_n \overrightarrow{r_{nb}}}{\mathrm{dt}} + \frac{\mathrm{d}_n \overrightarrow{r_{bd}}}{\mathrm{dt}} \tag{2.8}$$

The first term on the right of the above equation is the speed of movement between two coordinate systems, put Eq. (2.4) into the second term on the right, unfold it:

$$\frac{\mathrm{d}_n \overrightarrow{r_{bd}}}{\mathrm{dt}} = \frac{\mathrm{d}_n r_{bdx}}{\mathrm{dt}} \overrightarrow{i_b} + \frac{\mathrm{d}_n r_{bdy}}{\mathrm{dt}} \overrightarrow{j_b} + \frac{\mathrm{d}_n r_{bdz}}{\mathrm{dt}} \overrightarrow{k_b} + r_{bdx} \frac{\mathrm{d}_n \overrightarrow{i_b}}{\mathrm{dt}} + r_{bdy} \frac{\mathrm{d}_n \overrightarrow{j_b}}{\mathrm{dt}} + r_{bdz} \frac{\mathrm{d}_n \overrightarrow{k_b}}{\mathrm{dt}} \tag{2.9}$$

We can see the first three of the above equation as B and N systems have no relative motion, it means $\frac{\mathrm{d}_n \overrightarrow{r_{bd}}}{\mathrm{dt}} = \frac{\mathrm{d}_b \overrightarrow{r_{bd}}}{\mathrm{dt}}$ and $\frac{\mathrm{d}_n \overrightarrow{i_b}}{\mathrm{dt}} = \overrightarrow{\omega_{nb}} \times \overrightarrow{i_b}$, $\frac{\mathrm{d}_n \overrightarrow{j_b}}{\mathrm{dt}} = \overrightarrow{\omega_{nb}} \times \overrightarrow{j_b}$, $\frac{\mathrm{d}_n \overrightarrow{k_b}}{\mathrm{dt}} = \overrightarrow{\omega_{nb}} \times \overrightarrow{k_b}$ when $\overrightarrow{\omega_{nb}} = \overrightarrow{0}$. Combine the Eq. (2.4), we can write the Eq. (2.9) is:

$$\frac{\mathrm{d}_n \overrightarrow{r_{bd}}}{\mathrm{dt}} = \frac{\mathrm{d}_b \overrightarrow{r_{bd}}}{\mathrm{dt}} + \overrightarrow{\omega_{nb}} \times \overrightarrow{r_{bd}} \tag{2.10}$$

The above equation is Coriolis theorem in vector form, take it to the Eq. (2.8), we can get:

$$\frac{\mathrm{d}_n \overrightarrow{r_{nd}}}{\mathrm{dt}} = \frac{\mathrm{d}_n \overrightarrow{r_{nb}}}{\mathrm{dt}} + \frac{\mathrm{d}_b \overrightarrow{r_{bd}}}{\mathrm{dt}} + \overrightarrow{\omega_{nb}} \times \overrightarrow{r_{bd}} \tag{2.11}$$

Take the derivative of both sides of the above equation with respect to time in the N-system, and the vector composition relation between accelerations can be obtained:

$$\frac{\mathrm{d}_n^2 \overrightarrow{r_{nd}}}{\mathrm{dt}^2} = \frac{\mathrm{d}_n^2 \overrightarrow{r_{nb}}}{\mathrm{dt}^2} + \frac{\mathrm{d}_n}{\mathrm{dt}} \left(\frac{\mathrm{d}_b \overrightarrow{r_{bd}}}{\mathrm{dt}} \right) + \frac{\mathrm{d}_n}{\mathrm{dt}} \left(\overrightarrow{\omega_{nb}} \times \overrightarrow{r_{bd}} \right) \tag{2.12}$$

Applying Coriolis theorem to the second and third terms on the right of the above formula, it can be obtained:

$$\frac{\mathrm{d}_n}{\mathrm{dt}} \left(\overrightarrow{\omega_{nb}} \times \overrightarrow{r_{bd}} \right) = \frac{\mathrm{d}_b \overrightarrow{\omega_{nb}}}{\mathrm{dt}} \times \overrightarrow{r_{bd}} + \overrightarrow{\omega_{nb}} \times \frac{\mathrm{d}_b \overrightarrow{r_{bd}}}{\mathrm{dt}} + \overrightarrow{\omega_{nb}} \times \left(\overrightarrow{\omega_{nb}} \times \overrightarrow{r_{bd}} \right) \tag{2.13}$$

Because of $\frac{\mathrm{d}_n \overrightarrow{\omega_{nb}}}{\mathrm{dt}} = \frac{\mathrm{d}_b \overrightarrow{\omega_{nb}}}{\mathrm{dt}} + \overrightarrow{\omega_{nb}} \times \overrightarrow{\omega_{nb}}$ and $\overrightarrow{\omega_{nb}} \times \overrightarrow{\omega_{nb}} = \overrightarrow{0}$, $\frac{\mathrm{d}_n \overrightarrow{\omega_{nb}}}{\mathrm{dt}} = \frac{\mathrm{d}_b \overrightarrow{\omega_{nb}}}{\mathrm{dt}} = \frac{\mathrm{d} \overrightarrow{\omega_{nb}}}{\mathrm{dt}}$, take it to the Eq. (2.12), we can gain:

$$\frac{d_n^2\overrightarrow{r_{nd}}}{dt^2} = \frac{d_n^2\overrightarrow{r_{nb}}}{dt^2} + \frac{d_b^2\overrightarrow{r_{bd}}}{dt^2} + 2\overrightarrow{\omega_{nb}} \times \frac{d_b\overrightarrow{r_{bd}}}{dt} + \frac{d\overrightarrow{\omega_{nb}}}{dt} \times \overrightarrow{r_{bd}} + \overrightarrow{\omega_{nb}} \times \left(\overrightarrow{\omega_{nb}} \times \overrightarrow{r_{bd}}\right) \tag{2.14}$$

The formula above is the absolute acceleration vector synthesis formula, just for the sake of the derivation, can make N and B system origin coincide, meanwhile $\overrightarrow{r_{nd}} = \overrightarrow{r_{bd}} = \overrightarrow{r}, \overrightarrow{r_{nb}} = \overrightarrow{0}$, when $\overrightarrow{\omega_{nb}} = \overrightarrow{\Omega}$, Simplify the above equation further:

$$\overrightarrow{a_{nd}} = \overrightarrow{a_{bd}} + 2\overrightarrow{\Omega} \times \overrightarrow{v_{bd}} + \frac{d\overrightarrow{\Omega}}{dt} \times \overrightarrow{r} + \overrightarrow{\Omega} \times \left(\overrightarrow{\Omega} \times \overrightarrow{r}\right) \tag{2.15}$$

The formula above is Coriolis theorem, among them:

$\overrightarrow{a_{nd}}$ is the acceleration of particle d relative to the inertial coordinate system (N system), also known as the absolute acceleration.

$\overrightarrow{a_{bd}}$ is the acceleration of particle d relative to the shell coordinate system (B system), also known as the relative acceleration;

$2\overrightarrow{\Omega} \times \overrightarrow{v_{bd}}$ is Coriolis acceleration, $\overrightarrow{v_{bd}}$ is the velocity of particle d relative to the shell coordinate system.

$\overrightarrow{\Omega} \times \left(\overrightarrow{\Omega} \times \overrightarrow{r}\right)$ is implicated centripetal acceleration.

$\frac{d\overrightarrow{\Omega}}{dt} \times \overrightarrow{r}$ is implicated tangential acceleration.

In order to analyze the motion characteristics of silicon micromechanical gyroscope in detail, we can assumption:

$$\overrightarrow{a_{nd}} = a_x\overrightarrow{i_b} + a_y\overrightarrow{j_b} + a_z\overrightarrow{k_b},\ \overrightarrow{a_{bd}} = \ddot{x}\overrightarrow{i_b} + \ddot{y}\overrightarrow{j_b} + \ddot{z}\overrightarrow{k_b},$$
$$\overrightarrow{r} = x\overrightarrow{i_b} + y\overrightarrow{j_b} + z\overrightarrow{k_b},\ \overrightarrow{\Omega} = \Omega_x\overrightarrow{i_b} + \Omega_y\overrightarrow{j_b} + \Omega_z\overrightarrow{k_b}$$

Expand the Eq. (2.15)

$$\begin{bmatrix} a_x \\ a_y \\ a_z \end{bmatrix} = \begin{bmatrix} \ddot{x} \\ \ddot{y} \\ \ddot{z} \end{bmatrix} + \begin{bmatrix} 0 & -2\Omega_z & 2\Omega_y \\ 2\Omega_z & 0 & -2\Omega_x \\ -2\Omega_y & 2\Omega_x & 0 \end{bmatrix}\begin{bmatrix} \dot{x} \\ \dot{y} \\ \dot{z} \end{bmatrix} + \begin{bmatrix} -(\Omega_y^2 + \Omega_z^2) & \Omega_x\Omega_y - \dot{\Omega}_z & \Omega_x\Omega_z + \dot{\Omega}_y \\ \Omega_y\Omega_x + \dot{\Omega}_z & -(\Omega_x^2 + \Omega_z^2) & \Omega_y\Omega_z - \dot{\Omega}_x \\ \Omega_z\Omega_x - \dot{\Omega}_y & \Omega_z\Omega_y + \dot{\Omega}_x & -(\Omega_y^2 + \Omega_x^2) \end{bmatrix}\begin{bmatrix} x \\ y \\ z \end{bmatrix} \tag{2.16}$$

Consider the mechanical structure of the gyro as a shell coordinate system, the motion of a mass is regarded as rigid body motion, when you have angular velocity $\overrightarrow{\Omega}$ input about the z axis, the model can be substituted into the Eq. (2.16). Since the mass only moves in the x–y plane, the displacement, velocity and acceleration in the z direction can be ignored; because of angular velocity $\overrightarrow{\Omega}$ have component Ω_z in the z direction, we can simplify the Eq. (2.16):

$$\begin{bmatrix} a_x \\ a_y \end{bmatrix} = \begin{bmatrix} \ddot{x} \\ \ddot{y} \end{bmatrix} + \begin{bmatrix} 0 & -2\Omega_z \\ 2\Omega_z & 0 \end{bmatrix} \begin{bmatrix} \dot{x} \\ \dot{y} \end{bmatrix} + \begin{bmatrix} -\Omega_z^2 & -\dot{\Omega}_z \\ \dot{\Omega}_z & -\Omega_z^2 \end{bmatrix} \begin{bmatrix} x \\ y \end{bmatrix} \tag{2.17}$$

2.2.2 Silicon Based MEMS Gyroscope Dynamic Equation

The dual-mass linear vibration silicon-based MEMS gyroscope introduced in this book is a fully decoupled structure. Under ideal conditions, the motion between the driving mode and the sensing mode in its structure does not affect each other, Fig. 2.2 shows the diagram. The overall structure can be equivalent to two spring-mass-damping systems, respectively along the driving axis X and the sensing axis Y and the Z-axis is the sensitive axis, the dotted ring in the figure is the motion limiting structure. When it is in the working state, the driving mode drives the driving frame m_{xf} and the Coriolis mass block m_c to vibrate along the driving axis under the action of the driving force. When angular rate Ω_z is input around Z axis, m_c will be subjected to Coriolis force along y direction according to Coriolis theorem, which will drive sensing frame m_{yf} and m_c to move along Y direction. According to Newton's second law, the product of acceleration and mass of a mass block is the sum of external forces, so the force equation in the driving direction and detection direction is:

$$\begin{bmatrix} m_x a_x \\ m_y a_y \end{bmatrix} = \begin{bmatrix} F_{dx} \\ F_{dy} \end{bmatrix} - \begin{bmatrix} F_{kx} \\ F_{ky} \end{bmatrix} - \begin{bmatrix} F_{cx} \\ F_{cy} \end{bmatrix} + \begin{bmatrix} F_{ex} \\ F_{ey} \end{bmatrix} \tag{2.18}$$

In this equation, $m_x = m_{xf} + m_c$ is equivalent mass of driving mode; $m_y = m_{yf} + m_c$ is sensing mode equivalent mass; F_d, F_k, F_c, F_e is the matrix of electrostatic

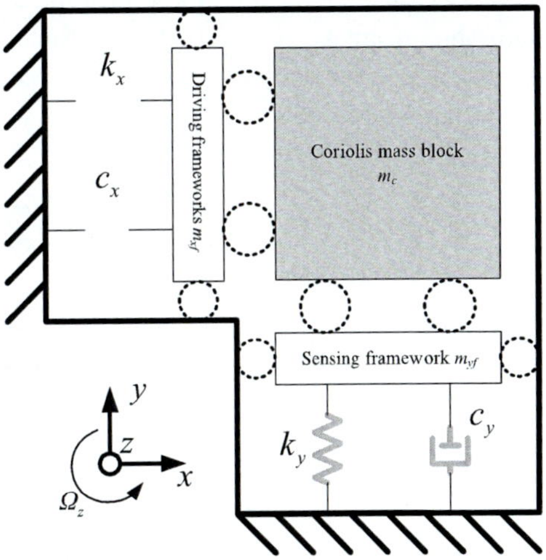

Fig. 2.2 Ideal mechanical model of dual-mass linear vibration silicon-based MEMS gyroscope structure

force, elastic force, damping force and system external force (e.g. shock, vibration, etc.). Under ideal conditions, ignoring the external force on the system, the elastic force and damping force matrix in the gyro structure can be obtained from Fig. 2.2:

$$\begin{bmatrix} F_{kx} \\ F_{ky} \end{bmatrix} = \begin{bmatrix} k_x & 0 \\ 0 & k_y \end{bmatrix} \begin{bmatrix} x \\ y \end{bmatrix} \begin{bmatrix} F_{cx} \\ F_{cy} \end{bmatrix} = \begin{bmatrix} c_x & 0 \\ 0 & c_y \end{bmatrix} \begin{bmatrix} \dot{x} \\ \dot{y} \end{bmatrix} \tag{2.19}$$

In this equation, k_x, c_x, k_y, c_y are the equivalent stiffness and damping of the driving mode and the sensing mode respectively; x and y are the displacement of driving and sensing directions, take the Eqs. (2.17), (2.19) into the Eq. (2.18), we can gained:

$$\begin{bmatrix} m_x\ddot{x} \\ m_y\ddot{y} \end{bmatrix} + \begin{bmatrix} c_x & -2m_x\Omega_z \\ 2m_y\Omega_z & c_y \end{bmatrix} \begin{bmatrix} \dot{x} \\ \dot{y} \end{bmatrix} + \begin{bmatrix} k_x - m_x\Omega_z^2 & -m_x\dot{\Omega}_z \\ m_y\dot{\Omega}_z & k_y - m_y\Omega_z^2 \end{bmatrix} \begin{bmatrix} x \\ y \end{bmatrix} = \begin{bmatrix} F_{dx} \\ F_{dy} \end{bmatrix} \tag{2.20}$$

As the mass block is assumed to be rigid body in the above section, however, in the x–y plane, the motion of m_{xf} in the structure is limited along the Y-axis, while the motion of m_{yf} is limited along the X-axis, so the force exerted on the rigid body only includes m_c, and the above equation should be adjusted as follows:

$$\begin{cases} m_x\ddot{x} + c_x\dot{x} + k_x x = F_{dx} + 2m_c\Omega_z\dot{y} + m_c\Omega_z^2 x + m_c\dot{\Omega}_z y \\ m_y\ddot{y} + c_y\dot{y} + k_y y = F_{dy} - 2m_c\Omega_z\dot{x} - m_c\dot{\Omega}_z x + m_c\Omega_z^2 y \end{cases} \tag{2.21}$$

In the above equation, because the displacement in the driving and detecting directions is very small, and $m_c\Omega_z^2 << k_x$, we can ignore $m_c\dot{\Omega}_z y$, $m_c\dot{\Omega}_z x$, $m_c\Omega_z^2 x$, $m_c\Omega_z^2 y$. And because the electrostatic driving force F_{dx} is very large in the driving mode, so $2m_c\Omega_z\dot{y}$ can be ignored (in the case of large detection displacement, this item should not be ignored). In addition, in the case of detecting open loop, the electrostatic force subjected to the detected mode $F_{dy} = 0$. The above equation can be further simplified as:

$$\begin{cases} m_x\ddot{x} + c_x\dot{x} + k_x x = F_{dx} \\ m_y\ddot{y} + c_y\dot{y} + k_y y = -2m_c\Omega_z\dot{x} \end{cases} \tag{2.22}$$

The equation above is the ideal state dynamics equation of the linear vibration silicon micromechanical gyroscope when detecting the open loop. The analysis in the following parts of this chapter is based on the sub-equation.

2.3 Silicon Based MEMS Gyroscope Motion Equation

In order to better design the measurement and control circuit of the gyroscope, it is necessary to analyze the motion characteristics of the two modes in its structure more directly.

$$\begin{cases} \ddot{x} + \dfrac{\omega_x}{Q_x}\dot{x} + \omega_x^2 x = \dfrac{F_{dx}}{m_x} \\ \ddot{y} + \dfrac{\omega_y}{Q_y}\dot{y} + \omega_y^2 y = -\dfrac{2m_c\Omega_z\dot{x}}{m_y} \end{cases} \tag{2.23}$$

$\omega_x = \sqrt{\frac{k_x}{m_x}}$, $\omega_y = \sqrt{\frac{k_y}{m_y}}$, $Q_x = \frac{m_x\omega_x}{c_x}$, $Q_y = \frac{m_y\omega_y}{c_y}$ are the resonant angular frequency and quality factor of the driving and sensing mode. It can be seen from the above formula that the proportion of m_c in m_y is directly proportional to the magnitude of Coriolis force, for analysis purposes, the $m_c = m_y$ here, assume that the electrostatic force subjected to the driving mode is a sine wave with constant frequency and constant amplitude, that is $F_{dx} = F_d\sin(\omega_d t)$, where F_d is its driving amplitude and ω_d is its driving angular frequency. The expressions of driving displacement and detection displacement can be obtained by substituting into Eq. (2.23):

$$\begin{aligned} x(t) = {} & \frac{F_d/m_x}{\sqrt{\left(\omega_x^2-\omega_d^2\right)^2+\omega_x^2\omega_d^2/Q_x^2}}\sin(\omega_d t+\varphi_x)+ \\ & \frac{F_d\omega_x\omega_d/m_xQ_x}{\left(\omega_x^2-\omega_d^2\right)^2+\omega_x^2\omega_d^2/Q_x^2}e^{-\frac{\omega_x}{2Q_x}t}\cos(\sqrt{1-1/4Q_x^2}\omega_x t)+ \\ & \frac{F_d\omega_d\left(\omega_x^2/Q_x^2+\omega_d^2-\omega_x^2\right)/m_x}{\omega_x\sqrt{1-1/4Q_x^2}\left[\left(\omega_x^2-\omega_d^2\right)^2+\omega_x^2\omega_d^2/Q_x^2\right]}e^{-\frac{\omega_x}{2Q_x}t}\sin(\sqrt{1-1/4Q_x^2}\omega_x t) \end{aligned} \tag{2.24}$$

$$\begin{aligned} y(t) = {} & \frac{F_c}{\sqrt{\left(\omega_y^2-\omega_d^2\right)^2+\omega_y^2\omega_d^2/Q_y^2}}\sin(\omega_d t+\varphi_x+\frac{\pi}{2}+\varphi_y)+ \\ & \frac{-F_c\left[\omega_y\omega_d\sin\varphi_x/Q_y+\left(\omega_y^2-\omega_d^2\right)\cos\varphi_x\right]}{\left(\omega_y^2-\omega_d^2\right)^2+\omega_y^2\omega_d^2/Q_y^2}e^{-\frac{\omega_y}{2Q_y}t}\cos(\sqrt{1-1/4Q_y^2}\omega_y t)+ \\ & \frac{F_c\left[\omega_y\left(\omega_y^2-3\omega_d^2\right)\cos\varphi_x/\left(2Q_y\right)+\omega_d\left(\omega_y^2/\left(2Q_y^2\right)+\omega_y^2-\omega_d^2\right)\sin\varphi_x\right]}{\omega_y\sqrt{1-1/4Q_x^2}\left[\left(\omega_y^2-\omega_d^2\right)^2+\omega_y^2\omega_d^2/Q_y^2\right]}\times \\ & e^{-\frac{\omega_y}{2Q_y}t}\sin(\sqrt{1-1/4Q_y^2}\omega_y t) \end{aligned} \tag{2.25}$$

Among them $\varphi_x = -tg^{-1}\left(\frac{\omega_x\omega_d}{Q_x(\omega_x^2-\omega_d^2)}\right)$, $F_c = \frac{-2\Omega_z\omega_d F_d}{m_x\sqrt{(\omega_x^2-\omega_d^2)^2+\omega_x^2\omega_d^2/Q_x^2}}$, $\varphi_y = -tg^{-1}\left(\frac{\omega_y\omega_d}{Q_y(\omega_y^2-\omega_d^2)}\right)$.

From the Eqs. (2.24) and (2.25), The motion of driving and sensing modes is the compound vibration of steady state vibration and attenuated vibration, so the displacement of driving and sensing modes in steady state can be regarded as simple harmonic vibration:

$$x(t) = \frac{F_d/m_x}{\sqrt{(\omega_x^2-\omega_d^2)^2+\omega_x^2\omega_d^2/Q_x^2}}\sin(\omega_d t+\varphi_x) \tag{2.26}$$

$$y(t) = \frac{F_c}{\sqrt{(\omega_y^2-\omega_d^2)^2+\omega_y^2\omega_d^2/Q_y^2}}\sin\left(\omega_d t+\varphi_x+\frac{\pi}{2}+\varphi_y\right) \tag{2.27}$$

It can be seen from Eq. (2.27) and F_c expression that F_c should be maximized in order to obtain the best detection displacement. Then Fc is maximized when $\omega_x = \omega_d$ and only when other variables remain unchanged, that is, the driving force frequency should be the resonant frequency of the driving mode, in which case $\varphi_x = 90°$ (or $-90°$). At the same time, in order to ensure better dynamic performance and bandwidth in engineering, the sensing and driving mode resonance frequency is usually kept at several kilohertz and the frequency difference of tens to hundreds of Hertz is retained. In addition, the form of vacuum encapsulation is usually used to make its quality factor in the range of several hundred to several thousand, then φ Y can be approximated as 0°. The above two equations are further simplified:

$$x(t) = \frac{F_d Q_x}{m_x\omega_d^2}\cos(\omega_d t) = A_x\cos(\omega_d t) \tag{2.28}$$

$$y(t) = \frac{-2\Omega_z F_d Q_x}{m_x\omega_d\sqrt{(\omega_y^2-\omega_d^2)^2+\omega_y^2\omega_d^2/Q_y^2}}\sin(\omega_d t) = A_y\sin(\omega_d t) \tag{2.29}$$

where, A_x and A_y are respectively the vibration amplitude of the driving and sensing modes, and the mechanical sensitivity of gyro structure can be obtained by further simplification of Eq. (2.29):

$$S_{mechanical} = \frac{A_y}{\Omega_z} \approx \frac{-F_d Q_x}{m_x\omega_d^2(\omega_y-\omega_d)} = \frac{-A_x}{\Delta\omega} \tag{2.30}$$

where $\Delta\omega$ is the difference between driving and sensing mode angular frequency. It can be seen from the above equation that the mechanical sensitivity of the structure is directly proportional to the vibration amplitude of the driving mode and inversely proportional to the frequency difference between the driving mode and the detection mode. At the same time, the small drive mass and low drive mode resonant frequency are beneficial to improve the mechanical sensitivity of gyro structure, but the low frequency ω_d is easily affected by external vibration. Therefore, in engineering, mechanical sensitivity is usually improved by reducing the drive mass m_x, frequency difference $\Delta\omega$ and increasing Q_x.

2.4 Comb Electrostatic Drive and Detection Principle

2.4.1 *Plate Capacitor Electrostatic Force Generation Principle*

When the voltage is applied to the capacitor plates of the parallel plate, the electrostatic force will be generated between the two plates, as shown in Fig. 2.3. In the figure, d_{0pad} is the distance between the two plates along the Y-axis. l_{pad} is the stacking length of the two plates in the X-axis direction. h_{pad} is the length of two plates in the z direction. Then the energy gradient of the electrostatic force vector can be expressed

$$\vec{F}_{pad} = -\nabla U_{pad} \tag{2.31}$$

Then the electric energy stored in the upper and lower plates is:

$$U_{pad} = \frac{Q_{pad}^2}{2C_{pad}(x, y, z)} = \frac{1}{2}C_{pad}(x, y, z)V_{pad}^2 \tag{2.32}$$

where, Q_{pad} is the charge stored in capacitor C_{pad}, and the potential energy of the system is:

$$U_{pad} = \frac{1}{2}C_{pad}(x, y, z)V_{pad}^2 - Q_{pad}V_{pad} = -\frac{1}{2}C_{pad}(x, y, z)V_{pad}^2 \tag{2.33}$$

Then the electrostatic force is:

$$\vec{F}_{pad} = -\nabla\left(-\frac{1}{2}C_{pad}(x, y, z)V_{pad}^2\right) = \frac{1}{2}\nabla C_{pad}(x, y, z)V_{pad}^2 \tag{2.34}$$

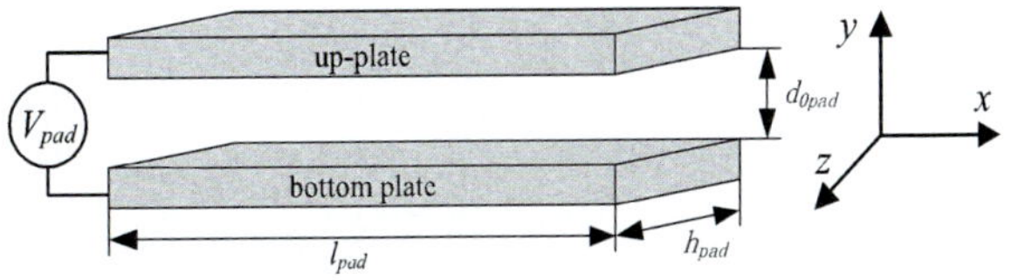

Fig. 2.3 Schematic diagram of electrostatic force generation

A capacitor

$$C_{pad}(x, y, z) = \varepsilon_0 \frac{h_{pad}(l_{pad} + x)}{d_{0pad} + y} \tag{2.35}$$

Among them, $\varepsilon_0 = 8.85 \times 10 - 12$F/m is dielectric constant. Substituting Eq. (2.34) into the above equation, we can get:

$$\begin{aligned} \vec{F}_{pad} &= \frac{1}{2}\frac{\partial C_{pad}(x, y, z)}{\partial x} V_{pad}^2 \hat{x} + \frac{1}{2}\frac{\partial C_{pad}(x, y, z)}{\partial y} V_{pad}^2 \hat{y} \\ &= \frac{\varepsilon_0 h_{pad} V_{pad}^2}{2}\left[\frac{\hat{x}}{d_{0pad} + y} - \frac{\hat{y}(l_{pad} + x)}{(d_{0pad} + y)^2}\right] \end{aligned} \tag{2.36}$$

where, $\hat{x}$ and $\hat{y}$ represent the vectors of forces along the x and y directions respectively.

2.4.2 *Comb Capacitance Detection Principle*

In order to increase the intensity of the detection signal, the gyroscope structure in this book adopts the comb capacitance detection form, which can transform the displacement signal into capacitance value for subsequent circuit extraction. According to the principle of variable area and space of plate capacitance, the capacitor of comb tooth can be divided into synovial film and pressure film correspondingly. Compared with the comb capacitance in the form of pressure film, there is a better linearity between the synovial capacitance and displacement, so in this book, both the driving test comb and test comb are in the form of synovial film. The structure diagram is shown in Fig. 2.4: there are movable comb teeth and the gyro mass driven by them in the middle, and fixed comb teeth on both sides. As a whole, the fixed comb teeth and moving comb teeth on the left can be regarded as the left capacitor C_{Lcomb}, and the fixed comb teeth and moving comb teeth on the right form the right capacitor C_{Rcomb}. According to the parameters in Fig. 2.4, the left and right capacitance can be expressed as

$$\begin{cases} C_{Lcomb} = n_{dcomb}\varepsilon_0 h_{comb}\left(\frac{l_{comb}+x}{d_0+y} + \frac{l_{comb}+x}{d_0-y}\right) + (2n_{dcomb}+1)\varepsilon_0 \frac{b_{comb}h_{comb}}{a_{comb}-x} \\ C_{Rcomb} = n_{dcomb}\varepsilon_0 h_{comb}\left(\frac{l_{comb}-x}{d_{0comb}+y} + \frac{l_{comb}-x}{d_{0comb}-y}\right) + (2n_{dcomb}+1)\varepsilon_0 \frac{b_{comb}h_{comb}}{a_{comb}+x} \end{cases} \tag{2.37}$$

where, n_{dcomb} is the number of comb teeth; l_{comb} is the stacking length of comb teeth; h_{comb} is the thickness of comb teeth; b_{comb} is the width of the comb tip; a_{comb} is the distance from the top of the comb tooth to the bottom of the opposite comb tooth; d_{0comb} is the parallel distance between comb teeth; x and y are the displacements of moving combs along x and Y axes respectively. Because the movable comb is restricted to move only along the x axis, then $y \ll d_0$, the above equation can be

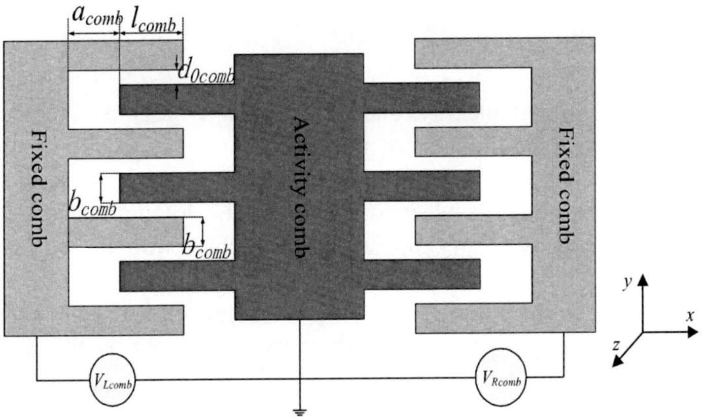

Fig. 2.4 Schematic diagram of comb tooth push–pull structure

further written as:

$$\begin{cases} C_{Lcomb} = 2n_{dcomb}\varepsilon_0 h_{comb}\left(\dfrac{l_{comb}+x}{d_0}\right) + (2n_{dcomb}+1)\varepsilon_0\dfrac{b_{comb}h_{comb}}{a_{comb}-x} \\ C_{Rcomb} = 2n_{dcomb}\varepsilon_0 h_{comb}\left(\dfrac{l_{comb}-x}{d_{0comb}}\right) + (2n_{dcomb}+1)\varepsilon_0\dfrac{b_{comb}h_{comb}}{a_{comb}+x} \end{cases} \tag{2.38}$$

In order to improve the signal-to-noise ratio and increase the detection signal amplitude, differential detection method is adopted in this book, so the capacitance variation obtained from the detection port is:

$$\begin{aligned} \Delta C_{comb} = C_{Lcomb} - C_{Rcomb} &= 4n_{dcomb}\varepsilon_0 h_{comb}\frac{x}{d_{0comb}} \\ &+ (2n_{dcomb}+1)\varepsilon_0 b_{comb}h_{comb}\left(\frac{2x}{a_{comb}^2 - x^2}\right) \end{aligned} \tag{2.39}$$

And when the driving displacement is not large, there is x ≪ a then the above formula can be further simplified to obtain:

$$\Delta C_{comb} \approx 4n_{dcomb}\varepsilon_0 h_{comb}\left(\frac{1}{d_{0comb}} + \frac{b_{comb}}{a_{comb}^2}\right)x \tag{2.40}$$

Through the above analysis, it can be seen that: under the condition of differential detection, the change of capacitor of synovial comb tooth is in a linear relationship with displacement, and the proportional coefficient is related to a series of structural parameters such as the number of comb teeth and the structural thickness. At the same time, smaller y displacement and x/a value are conducive to better linearity.

2.4.3 *Comb Capacitor Electrostatic Driving Principle*

Silicon structure has limited thickness and surface area. In order to increase the electrostatic driving force in the limited structural space, comb structure is usually used to increase the capacitance area, while the "push pull" force of bilateral symmetrical comb is used to increase the electrostatic force. The driving force and detection feedback force in this book adopt synovial comb structure, which is the same as the synovial detection comb described in the previous section. In order to facilitate the analysis of electrostatic force, the voltages applied to the left and right capacitors in Fig. 2.4 are V_{Lcomb} and V_{Rcomb} respectively, and these two voltages may be set as:

$$V_{Lcomb} = V_{dccomb} + V_{accomb}\sin(\omega_{comb}t) \tag{2.41}$$

$$V_{Rcomb} = V_{dccomb} - V_{accomb}\sin(\omega_{comb}t) \tag{2.42}$$

The V_{dccomb} and V_{accomb} are the amplitude of the DC and AC components of the driving voltage, ω_{comb} is the angular frequency that drives the AC component. After combining Eqs. (2.36) and (2.38) and ignoring the interfering force term, the electrostatic force in the x direction can be written:

$$F_{xcomb} = \frac{4n_{dcomb}\varepsilon_0 h_{comb} V_{dccomb} V_{accomb}\sin(\omega_{comb}t)}{d_{0comb}} \tag{2.43}$$

It can be seen from the above formula that the driving force of push–pull is proportional to the product of the number of comb teeth, the thickness of the structure and the dc and AC amplitude of the driving voltage, and inversely proportional to the distance between the parallel plates of the comb teeth. Under the condition of definite structure parameters, the electrostatic driving force can be adjusted in a larger range by adjusting the driving voltage. In conjunction with the equation of motion of the gyro structure described in the previous section, $\omega_{comb} = \omega_x = \omega_d$ is usually used to obtain the maximum amplitude of the drive mode.

2.5 Dual-Mass Linear Vibration Silicon-Based MEMS Gyroscope Structural Design

2.5.1 *Dual-Mass Linear Vibration Silicon-Based MEMS Gyroscope Structure Analysis*

Figure 2.5 is the schematic diagram of the sensitive structure of the dual-mass linear vibration silicon-based MEMS gyroscope studied in this book. It can be seen from the figure that the "spring" mentioned in Fig. 2.2 is the U-shaped girder in the structure above, which is more conducive to releasing the residual stress in the structure than the traditional straight beam. The drive and comb teeth of the gyroscope adopt the differential output synovium form, which can suppress the common mode interference and noise better, and improve the linearity and signal-to-noise ratio of the gyroscope output signal. At the same time, the symmetrical structure is used to reduce the influence of asymmetric factors and residual stress in the process of machining. In addition, the gyro structure adopts vacuum ceramic package, which can not only improve the quality factor of the structure, but also effectively reduce the influence of mechanical thermal noise on the signal. The left and right mass blocks are connected by connecting beam, which can make the left and right parts of the motion in the driving direction coupling each other, namely combined into a vibration system, which is more conducive to the subsequent circuit design. The structure adopts full decoupling scheme [71], so that the motion of driving and sensing modes do not affect each other. Because the stiffness of the u-beam mechanism along its two parallel long beam directions is very large, and the stiffness of the vertical direction is very small, u-beam can play a good role in motion isolation. Under the action of u-beam, the Coriolis mass block can move freely along x and y directions, while the driving frame and comb tooth can only move along x direction, while the movement of sensing frame and comb tooth is restricted in y axis direction. This scheme can greatly reduce the motion coupling between the driving and the sensing frame and reduce the nonlinear vibration of gyro. The structure used in this book has the following advantages [156]:

1. Fully decoupled structure design is adopted. The driving part and the sensing part have separate folding beams, which effectively reduces the mechanical coupling between the driving mode and the sensing mode.
2. The whole dual-mass gyro is symmetrically distributed in structure, and the detection differential output is formed by proper lead mode. It can not only eliminate the acceleration interference signal of the base along the detection axis, but also reduce the influence of temperature and other factors to the minimum through differential output, so as to improve the signal-to-noise ratio of the whole gyro.
3. The electrostatic comb capacitances and detection capacitances adopt the method of variable overlapping area, the movable comb tooth acts as synovial damping motion compared with the fixed comb tooth, the amplitude is larger, and the modal quality factor of the micro-gyro in the air is significantly improved. In addition, compared with the variable spacing method, the variable overlap area method has better linearity, and the dynamic range of measured angular rate is larger in the case of open loop detection.
4. The design of the beam uses folded beam form, so that the micro-gyro can work within the linear elastic deformation range of the beam, and the vibration is stable, and the use of a large number of folded beams can effectively reduce the residual stress introduced by machining.

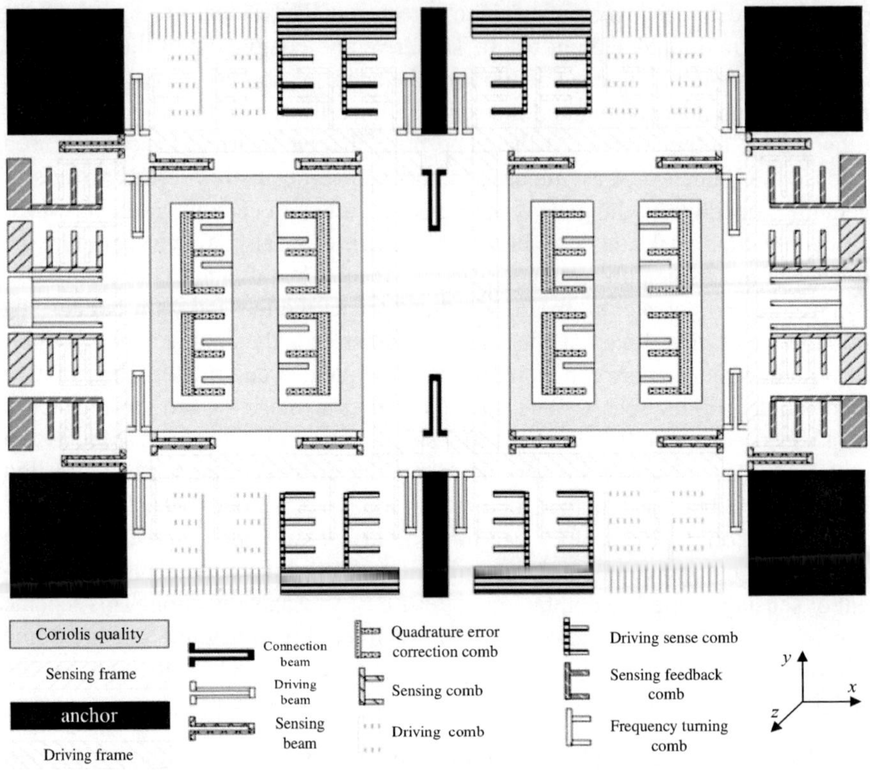

Fig. 2.5 Schematic diagram of sensitive structure of dual-mass linear vibration silicon-based MEMS gyroscope

5. The left and right mass blocks are connected by folded beams, which can reduce the interference between the left and right parts and effectively improve the signal-to-noise ratio of gyro.

According to the above advantages, the structural parameters are further designed as shown in Table 2.1:

2.5.2 *Structural Modal Analysis*

This book uses ANSYS, the finite element analysis software commonly used in micro-mechanical structure simulation, to analyze the structure. The main steps of modal simulation are as follows:

1. The actual symmetrically decoupled silicon micromechanical gyroscope is reasonably simplified.
2. Establish the geometric model of the structure.

Table 2.1 Structural design parameters of the dual-mass linear vibration silicon-based MEMS gyroscope

Parameter	Value	Parameter	Value
Left (right) driving mass m_x	1.2355 mg	Sensing mode long beam length l_{ls}	400 μm
Left (right) sensing mass m_y	1.1653 mg	Sensing mode short beam length l_{ss}	25 μm
Left (right) Coriolis mass m_c	1.1232 mg	Sensing mode beam width w_s	10 μm
Stacking length of comb teeth l	20 μm	Thickness of the structure h	60 μm
Driving mode long beam length l_{ld}	350 μm	The comb tooth width b	4 μm
Driving beam short beam length l_{sd}	40 μm	Comb top–bottom spacing a	20 μm
Driving (connect) beam width w_d	10 μm	Parallel spacing of comb teeth d_0	6 μm
Left (right) sensing frame area S_s	2000 * 150 μm^2	Number of left (right) drive combs n_d	400
Left (right) driving frame area S_d	2000 * 400 μm^2	Number of comb teeth left (right) n_s	400
Left (right) Coriolis mass area S_c	2000 * 4000 μm^2	Number of left (right) driving sense combs n_{ds}	400

3. Define structural unit type and material properties.
4. Divide the network and define boundary conditions.
5. Modal analysis module is used for modal simulation.
6. View the results and display the modal diagram.

In order to improve the solving speed, the structure can be simplified by ANSYS simulation. For example, the hole of quadrature nulling comb tooth formed on the mass block and the comb tooth structure can be equivalent to a certain mass for Modal equivalence. The first six order natural frequencies and modes of the structure can be obtained through simulation, as shown in Fig. 2.6 and Table 2.2: The first order modes of the micromechanical gyroscope structure (As shown in Fig. 2.6a). The left and right driving parts vibrate in the same frequency and phase along the driving shaft. The second mode (Fig. 2.6b) is the linear vibration of the left and right driving parts inverting along the driving shaft with the same frequency, which is also the resonance frequency of the actual work of the micro-gyro driving. The third mode (Fig. 2.6c) is the vibration of the in-phase line of the detected part along the y-axis of the same frequency, which is the interference mode of the microgyro. The fourth mode (Fig. 2.6d) is the vibration of the detection part along the y-axis of the same frequency inversion line, which is the sensing mode of the microgyro. The fifth (Fig. 2.6e) and sixth modes (Fig. 2.6f) are the in-phase and anti-phase motions of the microgyro along the z-axis, which are also interference modes.

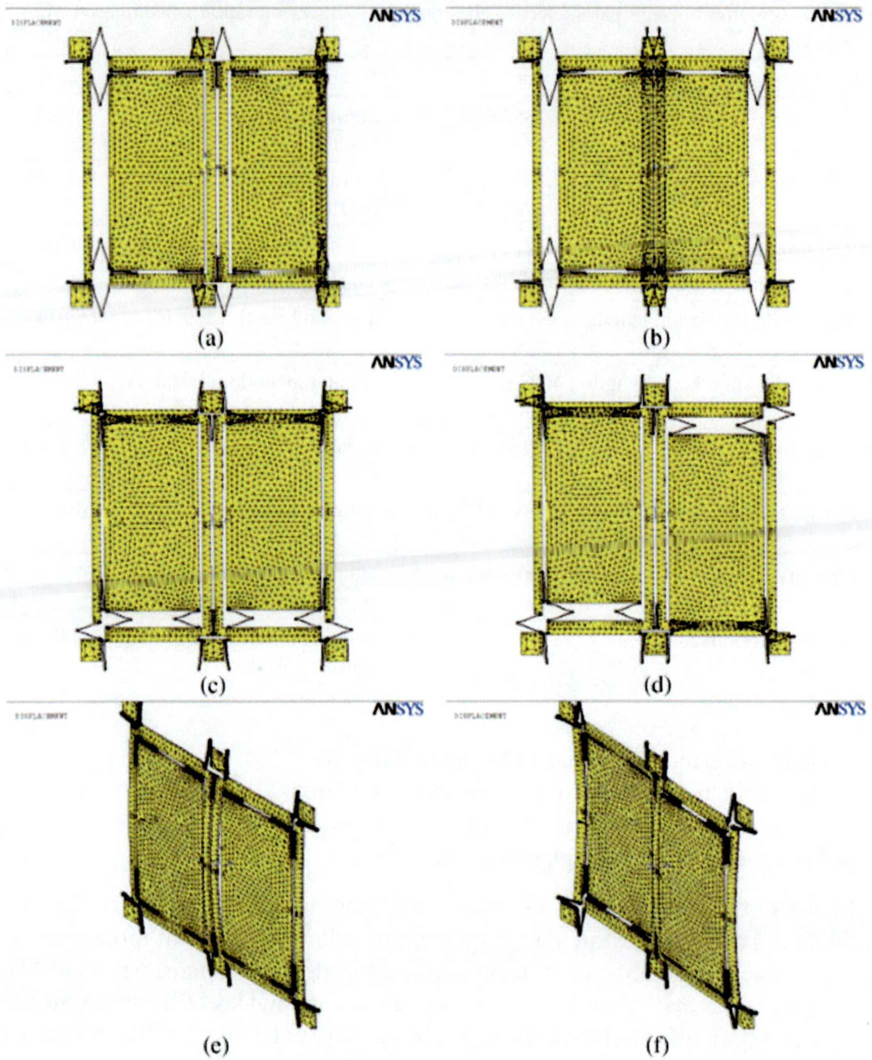

Fig. 2.6 Modal simulation diagram of dual-mass linear vibration silicon-based MEMS gyroscope

Table 2.2 First six modes of gyro structure

The modal number	1	2	3	4	5	6
Modal frequency (Hz)	2640	3095	3318	3441	4078	4793

When the micromechanical gyroscope working, the left and right parts are the line vibration of the same frequency and reverse phase. When the z-axis has angular velocity input, the left and right detection parts will only make the line vibration of the same frequency and reverse phase along the y-axis, that is, the detection mode works in the fourth mode. By adjusting the stiffness of the modes, the frequency difference between in-phase vibration and out-of-phase vibration can be adjusted along the drive shaft. In structural design, the frequency difference should be as large as possible to reduce the mutual interference between the two modes. When the natural frequencies of the driving mode and the sensing mode are ideally equal, the sensitivity of the microgyro is the highest, so the driving and sensing frequencies should be as close as possible in the design. In the above simulation results, the difference between the natural frequencies of the two modes is only 346Hz, and the frequencies of the two modes are well matched, and the frequency value is about 3200 Hz. In addition, the natural frequencies of the working modes are isolated from other interference modes to a certain extent. It can be seen from the driving mode diagram that when the two mass blocks of the microgyro are driving along the x-axis in the same frequency, the comb tooth and frame of the driving mode on the left and right are consistent with the motion of the mass block, while the comb tooth and frame of the sensing mode are basically static, that is, the driving and sensing modes are decoupled during the driving mode works. Similarly, it can also be seen from the sensing mode diagram that when the two mass blocks of the microgyro move in the same frequency and reverse phase along the y-axis, the comb tooth and frame of the left and right sensing mode move in the same way as the mass block, while the comb tooth and frame of the driving mode remain basically static, that is, the driving and sensing modes are decoupled during the sensing mode works. The results of modal simulation show that the structure of dual-mass linear vibration silicon-based MEMS gyroscope achieves complete structural decoupling follows the design purpose.

2.5.3 Static Stress Simulation Under Gravity

The vibration structure of the dual-mass linear vibration silicon-based MEMS gyroscope is supported on the anchor point by U-beams. The gyroscope structure handles only gravity when the gyroscope structure in static state. The gravity influence to the structure can be simulated and analyzed by loading 1 g gravity along x, y and z axes respectively. The three directions stress distribution as shown in Fig. 2.7 (a, b and c are the gravity stress distribution along x, y and z axes respectively), and the maximum stress points are shown in Fig. 2.8 (a, b and c are the maximum stress points under gravity action along x, y and z axes respectively). The maximum stresses all occur at the joints of beams, and the maximum values are shown in Table 2.3.

As can be seen from the simulation value, in the whole free state, the maximum stress value in three directions under the influence of gravity is very small, far lower than the ultimate strength value of 790 MPa for the fracture of monocrystalline silicon

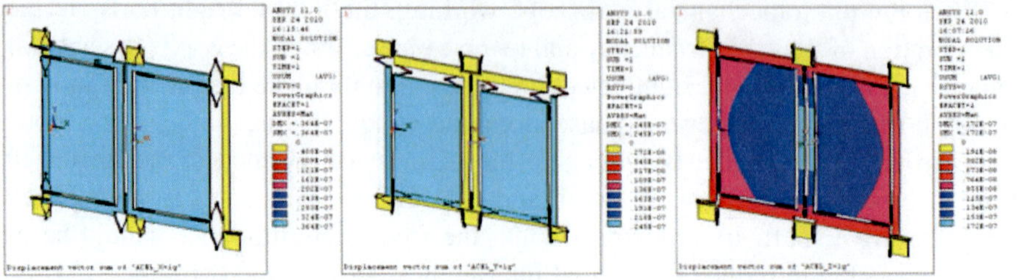

Fig. 2.7 Stress distribution under gravity in dual mass linear vibration silicon-based MEMS gyroscope

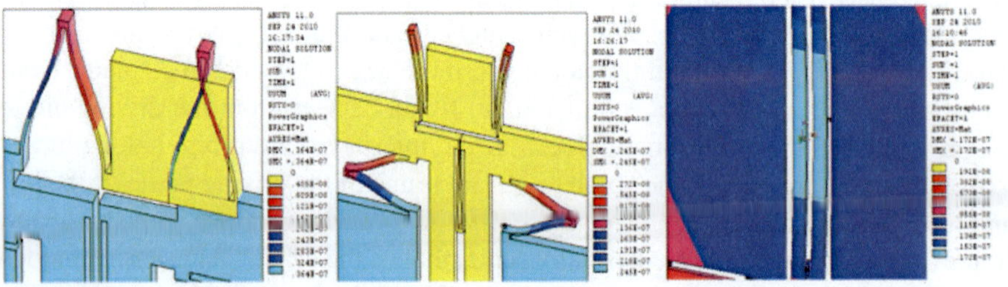

Fig. 2.8 Position of maximum stress under gravity in dual-mass linear vibration silicon-based MEMS gyroscope

Table 2.3 Maximum stress under gravity

Direction	x	y	z
Stress value (MPa)	0.0364	0.0245	0.0172

material. Therefore, in the free state, the strength of the entire silicon microgyro meets the requirements.

2.5.4 *Stress at Maximum Displacement*

Due to the unable-estimate factors in MEMS gyroscope structure, such as damping, residual stress and so on, it is difficult to reflect the electrostatic force and displacement in the ANSYS environment. At the same time, even if part of the movement to limit position moment in simulation load of electrostatic force is very difficult, so it is needed to adopt the method of applying displacement to load the finite element model. A displacement equivalent to the amplitude is applied to the driving part of the gyroscope in place of the electrostatic load. Because of the large number of combs and the symmetrical distribution of electrostatic field, it is feasible to replace the actual electrostatic load with displacement. In driving and sensing direction to load, driving part of the motion amplitude is determined by driving voltage amplitude, in simulation part, assuming the driving amplitude is 10 μm, because left and right driving parts achieve reverse movement, so when the simulation need to reverse the part about loading at the same time, to simulate a silicon micro gyroscope actual situation at work. In the detection state, the motion amplitude caused by Coriolis force is relatively small, so the value of limit state can be used for simulation. In the simulation, the displacement in the sensing mode direction is also 10 μm, and the load direction of the displacement of the left and right parts is opposite in the sensing mode direction. Simulation results are shown in Figs. 2.9 and 2.10. The maximum value also appears at the connection of the beam, and the maximum value of the driving direction and sensing direction is 102 MPa, which is far lower than the limit of fatigue fracture of silicon material.

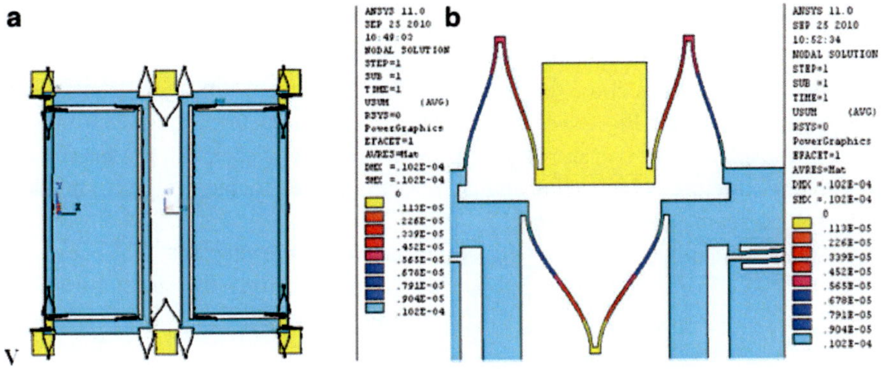

Fig. 2.9 Stress distribution (**a**) and position of maximum stress (**b**) at maximum displacement in driving direction

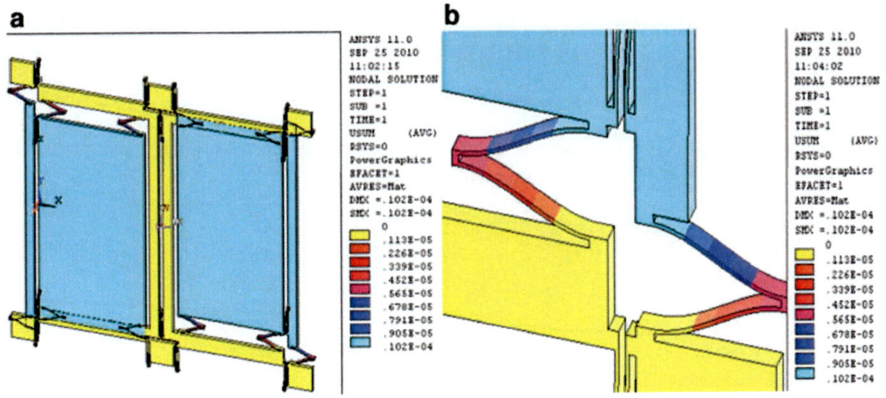

Fig. 2.10 Stress distribution (**a**) and position of maximum stress (**b**) at maximum displacement in sensing direction

2.5.5 Impact Simulation

The transient analysis module of ANSYS was used to simulate the impact of the dual-mass linear vibration silicon-based MEMS gyroscope structure, and the impact response results were analyzed at the geometric center point of the sensitive mass block on the left side of the model. Firstly, a sinusoidal impact signal with a peak value of 100 g and an action time of 6 ms was applied to the micromechanical gyroscope structure in the x, y and z directions to simulate the impact of the structure falling. Through the simulation results, it can be seen that:

In the impact on x direction, the maximum displacement occurs at 2.98 ms, 3.75 μm (as shown in Fig. 2.11a). By investigating the structural stress distribution under this displacement (As shown in Fig. 2.11b), it can be seen that the maximum stress of the structure is 39.8 MPa, and the maximum stress occurs at the connection of the driving folding beam (as shown in Fig. 2.11c).

In the impact on y direction, the maximum displacement occurs at 2.90 ms, which is 2.23 μm (as shown in Fig. 2.12a). By investigating the structural stress distribution under this displacement (As shown in Fig. 2.12b), it can be seen that the maximum stress of the structure is 29.4 MPa, and the maximum stress occurs at the connection of the folded beam (as shown in Fig. 2.12c).

In the case of impact on the Z direction, the maximum displacement occurs at 3.00 ms, 1.06 μm (as shown in Fig. 2.13a). By investigating the structural stress

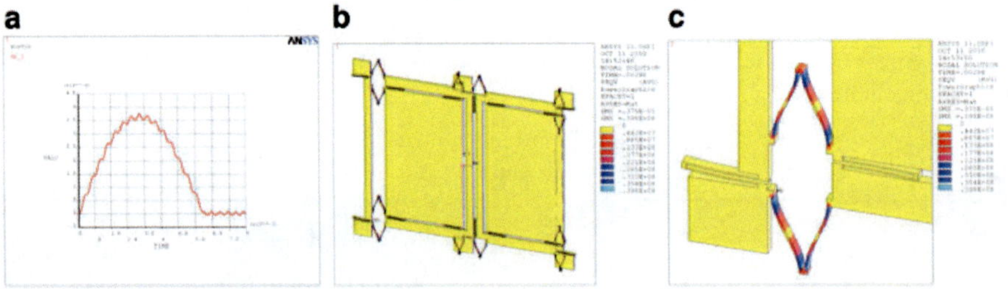

Fig. 2.11 100 g impact response curve in x direction (**a**) and stress distribution (**b**) and maximum stress position (**c**) at the time of maximum displacement

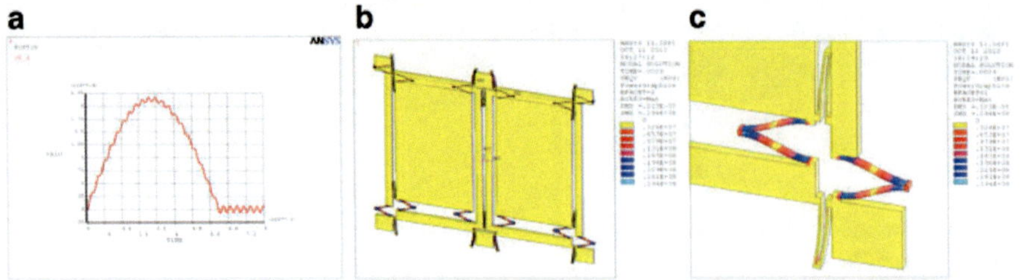

Fig. 2.12 100 g impact response curve in y direction (**a**) and stress distribution (**b**) and maximum stress position (**c**) at the time of maximum displacement

distribution under this displacement (As shown in Fig. 2.13b), it can be seen that the maximum stress of the structure is 28.5 MPa, and the maximum stress occurs at the connection of the driving folding beam (as shown in Fig. 2.13c). The analysis of the above simulation results shows that the maximum stress in three directions is far lower than the limit stress of the structural material of 790 MPa, which indicates that the structure is safe under the effect of drop impact.

Secondly, to test the biggest impact that the structure can withstand, the impact of more than 1000 g was applyed in three directions respectively, the maximum stress curves are shown in Fig. 2.14. It can be seen that the z direction achieves the worst shock resistance, x direction second, y direction shock resistance is best, the whole structure in three directions can be resistant over than 2000 g. However, in the impact simulation, the displacement of the moving part of the micromechanical gyroscope structure will exceed the limit position, so it is necessary to design a stop block to limit the structure design.

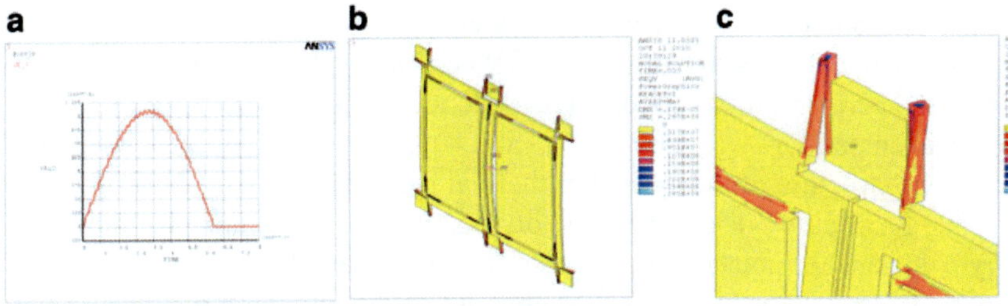

Fig. 2.13 100 g impact response curve in z direction (**a**) and stress distribution (**b**) and maximum stress position (**c**) at the time of maximum displacement

Fig. 2.14 Simulation curves of limit impact in three directions

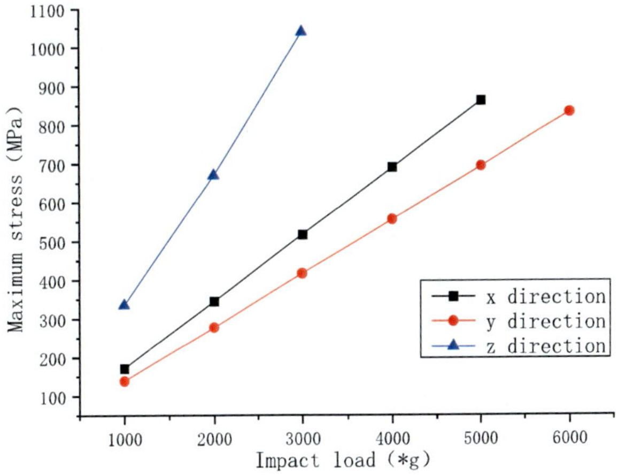

2.5.6 Thermal Stress Simulation

In dual-mass linear vibration silicon-based MEMS gyroscope, the silicon structure layer and the glass substrate are sealed together by electrostatic bonding technology. The bonded glass is 7740 glasses, which has a thermal expansion coefficient of 3.25 $\times$ 10^{-6}/K, and monocrystalline silicon has a thermal expansion coefficient of 2.6 $\times$ 10^{-6}/K. Electrostatic bonding is usually performed at a high temperature of 400 °C and cooled to room temperature after bonding. Due to the different thermal expansion coefficient of the two materials, a large mismatched thermal stress is generated at the bonding point, which is transferred to the vibration beam connected with the base. In order to analyze the effect of electrostatic bonding stress on resonant frequency, the thermal analysis of dual mass gyro was carried out.

Since the stress is mainly caused by the difference in thermal expansion coefficient between silicon and glass during bonding, in addition to the silicon structure, the structure of the glass substrate connected to the silicon base should be established during the establishment of the geometric model, as shown in Fig. 2.15a, where the height of the anchor point is 60 μm and the thickness of the glass substrate is 100 μm. The overall temperature of the silicon-glass structure was set at 400 °C (673 K), and the reference temperature was set at 27 °C (300 K). Figure 2.15b is the deformation diagram of the structure of the gyro after thermal analysis (temperature drops from 673 K to 300 K). Figure 2.15c is the stress distribution and maximum position of the micro-gyro and anchor points. It can be seen from the figure that the base connected to the glass deforms after thermal expansion and cold contraction due to different thermal expansion coefficients of silicon and glass. The maximum stress generated here is 491 MPa, and the maximum value is lower than the limit stress of silicon material 790 MPa. Therefore, the analysis results show that the whole micro-gyro can withstand the thermal stress caused by temperature change.

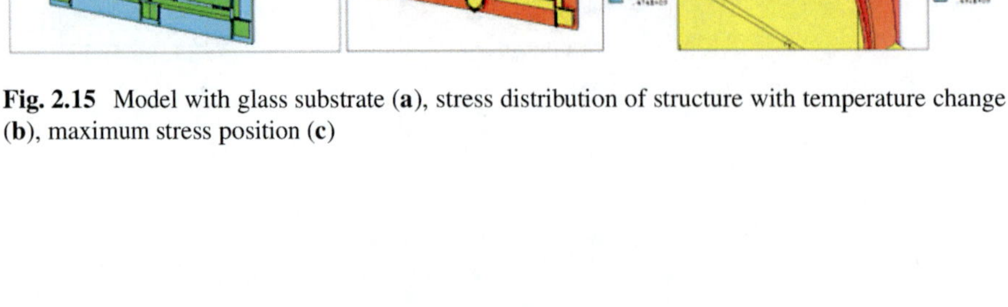

Fig. 2.15 Model with glass substrate (**a**), stress distribution of structure with temperature change (**b**), maximum stress position (**c**)

2.5.7 Harmonic Response Analysis

Any sustained periodic load will produce a sustained periodic response (harmonic response) in a linear structural system. Harmonic response analysis is a technique used to determine the steady-state response of a linear structure subjected to a sinusoidal gauge varying over time. By using ANSYS software to analyze the harmonic response of the model, the structural response of the structure under the sinusoidal load with known frequency can be determined. Through the harmonic response analysis, the response of the structure at a certain frequency point can be obtained, and also the response of the structure at a certain point in the whole frequency segment can be obtained. The main purpose of harmonic response analysis of silicon microgyroscope is to calculate the displacement response of the structure under the action of electrostatic force and obtain the amplitude-frequency response curve of silicon micromechanical gyroscope. In the simulation process, the point of applying the electrostatic force is the center point of the two mass blocks, the center point of the left mass block is point a, and the center point of the right mass block is point b.

First of all, to stimulate the driving in-phase vibration modal (a first order modal), harmonic response analysis in a, b, a driving force of the point on the x direction in phase (as shown in Fig. 2.16a), within the scope of driving mode to sweep frequency (2500–3500 Hz), a, b point displacement can be the only one of the biggest peak points, namely the first modal resonance frequency points. It shows that the model has only one in-phase vibration mode and the amplitudes of the two mass blocks are the same. In this way, the co-vibration mode is the main driving mode of the whole gyro. In dual-mass linear vibration silicon-based MEMS gyroscope, the driving inphase mode should be suppressed, which is not suitable for the driving circuit.

Second, to find the driving anti-phase vibration modal (second order mode), harmonic response analysis in a, b applying reverse driving force of the x direction (as shown in Fig. 2.17a), within the scope of drive mode to sweep frequency (2500–3500 Hz), a, b of displacement can be the only one of the biggest peak points, namely the second modal resonance frequency points. This indicates that the model has only one reverse vibration mode and the amplitudes of the two mass blocks are the same. In this way, the reverse vibration mode is the main driving mode of the

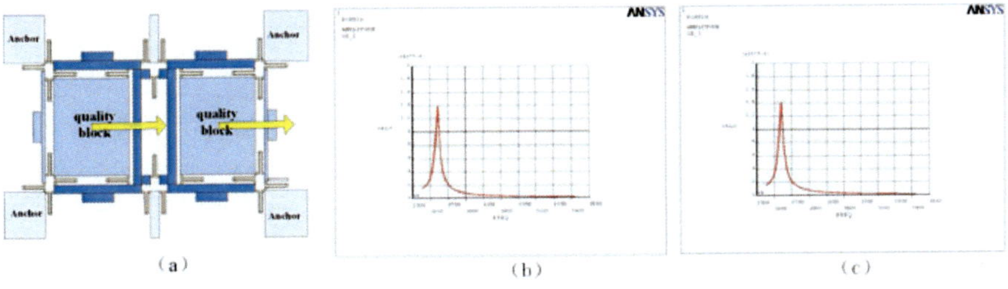

Fig. 2.16 Driving in-phase mode loading mode (**a**), amplitude-frequency characteristic diagram of point A (**b**), amplitude-frequency characteristic diagram of point B (**c**)

whole gyro. In dual-mass silicon microgyro, the driving anti-phase is the working mode of gyro and its drive circuit.

Again, to analyse sensing in-phase modal (third order modal), harmonic response analysis in a, applying a driving force of the y direction in phase b (as shown in Fig. 2.18a), within the scope of testing modal (3200–3600 Hz) to sweep, a, b of displacement can be the only one of the biggest peak points, namely the third modal resonance frequency points. It shows that the model has only one in phase vibration mode and the amplitudes of the two mass blocks are the same. In this way, the co-vibration mode is the main detection mode of the whole gyro. In dual-mass silicon microgyro, in-phase detection is the interference mode of gyro detection and should be suppressed.

Finally, to investigate the sensing anti-phase modal (fourth order modal), harmonic response analysis in a, b applying reverse driving force of the y direction (as shown in Fig. 2.19a), within the scope of testing modal (3200–3600 Hz) to sweep, a, b point displacement can be the only one of the biggest peak points, namely the fourth mode resonance frequency points. This indicates that the model has only one reverse vibration mode and the amplitudes of the two mass blocks are the same. When this loading mode is adopted, the sensing anti-phase mode is the detection mode of gyro, and the circuit design of the detection loop should be carried out around this mode.

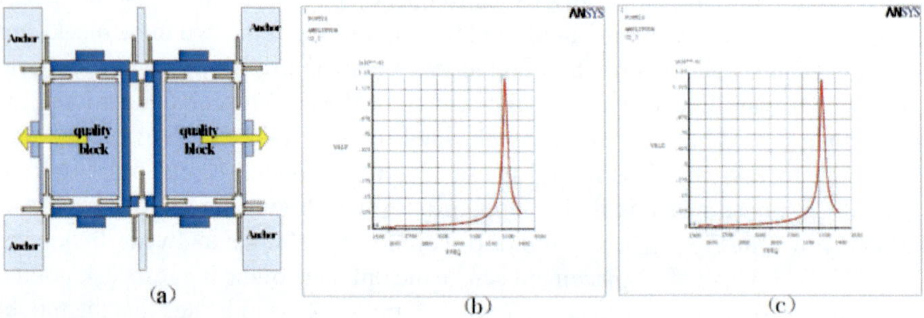

Fig. 2.17 Driving anti-phase mode loading mode (**a**), amplitude-frequency characteristic diagram of point A (**b**), amplitude-frequency characteristic diagram of point B (**c**)

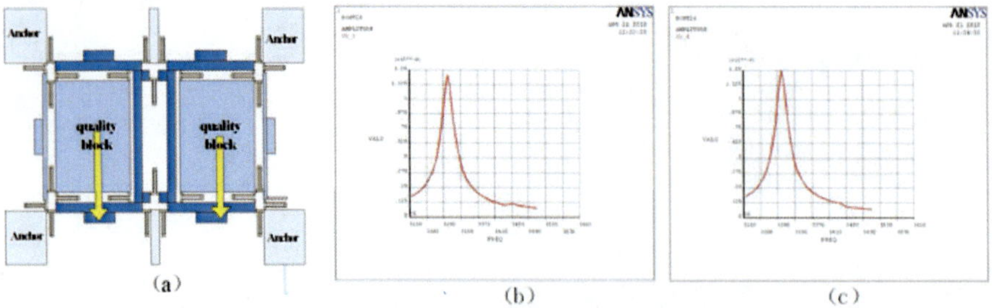

Fig. 2.18 Sensing in-phase modal loading mode (**a**), amplitude-frequency characteristic diagram of point A (**b**), amplitude-frequency characteristic diagram of point B (**c**)

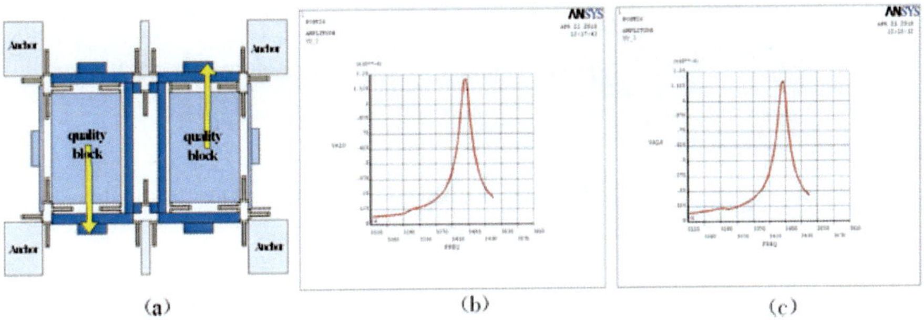

Fig. 2.19 Sensing anti-phase mode loading mode (**a**), amplitude-frequency characteristic diagram of point a (**b**), amplitude-frequency characteristic diagram of point b (**c**)

2.6 Silicon Based MEMS Gyroscope Structure Processing

Silicon micromechanical gyroscope structure processing adopts silicon-based micromachining technology, mainly including bulk silicon micromachining technology, surface silicon micromachining technology and composite micromachining technology, the above three technologies have their own scope of application. This book mainly introduces the current DDSOG (Deep Dry Silicon on Glass) process and Silicon on Insulator (SOI) process.

DDSOG process is the bonding of silicon materials and glass materials, and with dry deep silicon etching of silicon materials, can be processed into 100-micron thick silicon structure, the distance between the silicon structure and the glass substrate is large, the parasitic capacitance is small, the use of monocrystalline silicon raw material can obtain small structural residual stress characteristics. The process is suitable for processing thick and large mass block structures. The typical process is shown in Fig. 2.20: Firstly, the mask layer is graphically painted on the back of the silicon wafer, and a certain depth is etched by dry process, which ensures the distance between the final silicon wafer and the glass substrate. At the same time, a shallow groove is carved on the glass substrate, and the metal deposition into the shallow groove, forming a lead layer on the glass substrate; Then the silicon wafer back and the glass substrate bonding, so that the silicon structure and the corresponding metal lead plate completely corresponding to fit; Thirdly, the silicon wafer is cut to the desired thickness, which determines the final thickness of the silicon structure by subtracted from the thickness carved on the back; Finally, the silicon structure is released by deep silicon dry etching process on the front side of the silicon wafer.

SOI process uses an all-silicon structure, with a silicon dioxide layer interspersed in the middle. The silicon layer above the silicon dioxide layer (hereinafter referred to as the structural layer) is used to etch the silicon structure, and the silicon layer below the silicon dioxide layer (hereinafter referred to as the base layer) is used to provide structural support, similar to the glass substrate in DDSOG process. Because the structure and the substrate are silicon structure, the thermal expansion coefficient and other parameters are the same, so it can provide good temperature characteristics

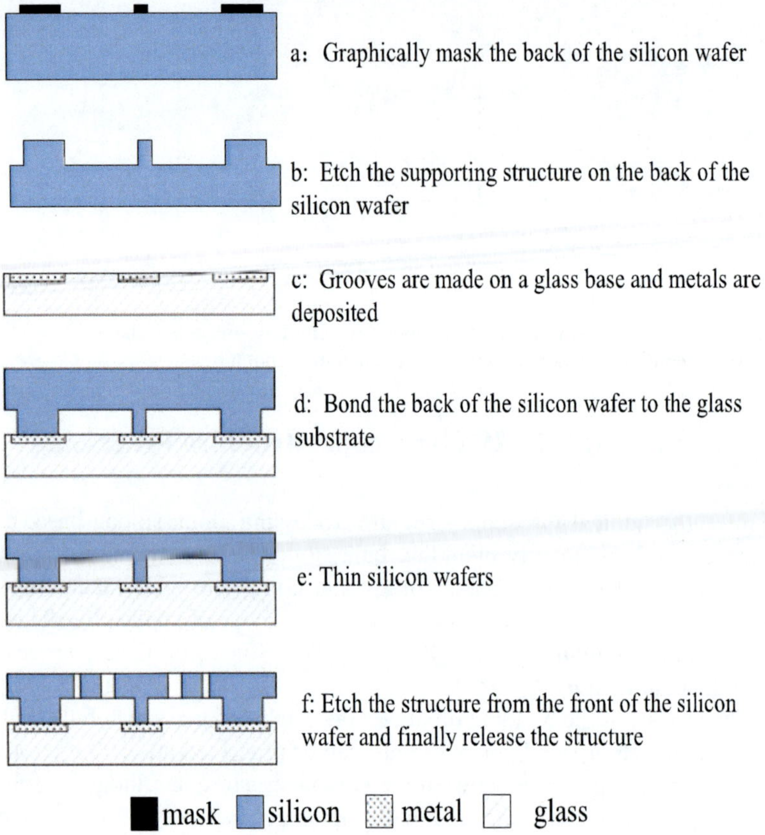

Fig. 2.20 Typical DDSOG process flowchart

for the micromechanical structure, the residual stress in DDSOG process can be alleviated in SOI process. But limited by the thickness of the middle silica layer, the distance between the structural layer and the base layer is only a few microns, which will produce a large parasitic capacitance. The typical process is shown in Fig. 2.21 [161]: First, a graphically mask layer is made on the surface of the silicon structural layer, and a certain depth is etched to reach the silicon dioxide layer by the dry process, which is the thickness of the silicon structure. Moreover, most of the structural layer and the silicon dioxide layer are further separated by over etching. Then, remove the mask layer; Thirdly, wet etching removes the silica layer. And finally, bonding metal leads.

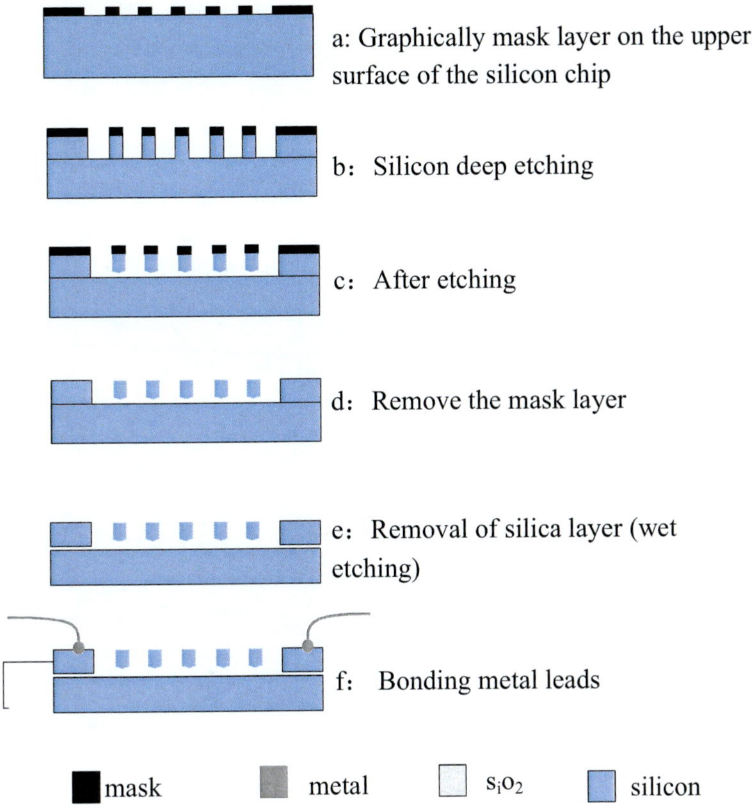

Fig. 2.21 Typical SOI process flowchart

2.7 Chapter Summary

This chapter first introduces the principle of Coriolis effect, and deduces its dynamic equation and motion equation based on the ideal mechanical model of dual-mass linear vibration silicon-based MEMS gyroscope, and analyzes the displacement characteristics and mechanical sensitivity of gyroscope structure driving and sensing mode. Secondly, capacitance detection and driving mechanism of synovial combs are introduced. Thirdly, the dual-mass linear vibration silicon-based MEMS gyroscope structure involved in this book is introduced, and the schematic diagram of the whole structure including quadrature correction, sensing force feedback comb and tuned comb are given, and the working principle of the fully decoupled structure and the principle of double mass block to suppress the overall acceleration in the detection direction are analyzed. At the same time, the simulation process and results of gyro structure in ANSYS software are introduced. Finally, combined with the common DDSOG and SOI processes, the processing method of the dual-mass linear vibration silicon-based MEMS gyroscope structure is introduced.

Chapter 3
Dual-Mass Linear Vibration Silicon-Based MEMS Gyroscope Structure Noise Analysis and System Model

3.1 Chapter Introduction

With the in-depth development of silicon micromechanical gyroscope related technology, its structural noise has gradually become one of the main factors affecting its accuracy [106]. Therefore, it is particularly important to conduct objective analysis of its structural noise and design the relevant high SNR (signal noise ratio) interface circuit. In addition, the establishment and improvement of the silicon micromechanical gyroscope structure system model containing major unsatisfactory factors can simulate the influence of each factor on the gyroscope performance, which is helpful to verify the control method of each loop and shorten the development and design cycle of gyroscope. In addition, valuable feedback information can be provided for the design of the gyroscope structure to further improve the performance of the gyroscope.

In this chapter, the equivalent electrical model of dual-mass linear vibration silicon-based MEMS gyroscope structure is established, and the interface circuit based on carrier modulation and ring diode demodulation is analyzed. Secondly, combined with the motion equation of the ideal gyroscope introduced in Chap. 2, the ideal gyroscope structure model is established in Simulink simulation environment, and the mechanical thermal noise term is introduced as the non-ideal factor. Thirdly, based on the automatic gain control (AGC) technology, the self-excited oscillation driving closed-loop of silicon micromechanical gyroscope is designed. The operating point and stability of the driving closed-loop system were analyzed, and the system simulation was carried out to verify the driving closed-loop system combined with the gyro structure model. Finally, the noise related to the dual-mass linear vibration silicon-based MEMS gyroscope structure is analyzed, and experiments are designed and tested according to different noise characteristics. The gyro structure model is also used as the basic model for the study of quadrature error correction technology, sensing closed-loop technology and frequency tuning technology in the following chapters, providing an accurate model simulation platform for the further in-depth study of dual-mass linear vibration silicon-based MEMS gyroscope.

H. Cao, *Dual-Mass Linear Vibration Silicon-Based MEMS Gyroscope*,
https://doi.org/10.1007/978-981-19-9247-6_3

3.2 Silicon Based MEMS Gyroscope Structure Equivalent Electrical Model and Interface

Although the gyro structure is mainly mechanical movement, displacement detection and electrostatic force generation are realized by capacitance. Therefore, the first task of obtaining high SNR interface is to establish accurate electrical model of gyro structure.

3.2.1 Silicon Based MEMS Gyroscope Structure Equivalent Electrical Model

This book is improved on the basis of previous work [70, 117]. In order to facilitate the introduction and analysis of the working principle of the interface circuit, the equivalent electrical model of dual mass linear vibration silicon-based MEMS gyroscope structure of the left mass block and its interface circuit are studied in this section (the electrical model of the right mass block is the same), as shown in Fig. 3.1. In the figure, the gray box is equivalent electrical model of gyro structure, where the equivalent nodes of mass block and comb tooth frame in silicon structure are represented by bold lines in different colors, and the colors of each part are corresponding to Fig. 2.5. The capacitors represented by the thick solid line are the working capacitors, where C_{dL+} and C_{dL-} are the driving capacitors. C_{dcL+} and C_{dcL-} are the driving detection capacitors. C_{sL+} and C_{sL-} are detection capacitors; R_{spr} is the resistance of U-beam, usually several hundred Ω; C_{dis} is the distributed capacitance between wires, usually several pF; R_G is the gold wire for the structure leading to the package pin, which is a few ohms; R_p and C_p are the parasitic resistances (about several hundred MΩ) and parasitic capacitors clamped at both ends of each working capacitor. The ground symbol (GLA symbol) in the gray box is connected to the glass base and the external circuit. The driving sense interface and sensing interface adopt the same circuit, which is composed of carrier source, precision capacitor, ring diode and instrument amplifier. In the figure above, V_{sL} and V_{dcL} serve as the pre-amplification output signals of the left mass block detection capacitor and the driver detection capacitor respectively. The anti-noise characteristics and working principle of the interface circuit will be described in detail in the following two sections.

3.2.2 Ring Diode Interface Circuit Noise Resisting Characteristic

For conventional charge amplifiers without high frequency carrier modulation, the spurious resistors R_{ps+} and R_{ps-} generate large scintillation noise and electronic noise

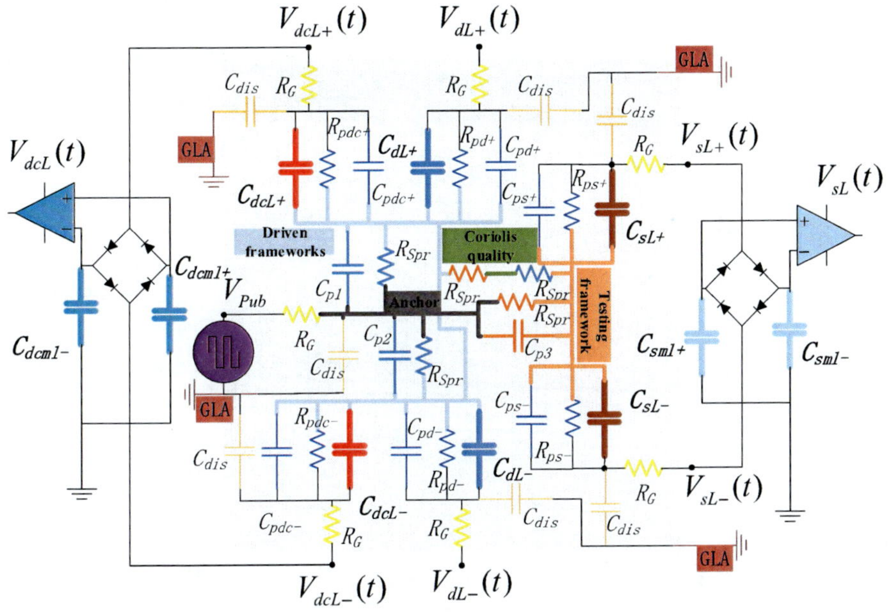

Fig. 3.1 Equivalent electrical model of dual-mass linear vibration silicon-based MEMS gyroscope left mass and interface circuit

due to their mego Ω level and the frequency of current flowing through them being the driving mode frequency (approximately several kHz) [111]. This book adopts the differential detection method of carrier generator and ring diode circuit to complete the detection of comb capacitance [120]. Since the R_{spr} is only a few hundred Ω, the circuit in Fig. 3.1 can be simplified (taking the interface of the sensing loop as an example), as shown in Fig. 3.2. In the figure, $V_{no+} = V_{ELENps+} + V_{FNps+}$ and $V_{no-} = V_{ELENps-} + V_{FNps-}$ are the superposition of the equivalent voltage of the electronic noise and scintillation noise generated by the parasitic resistance respectively. Firstly, the attenuation of scintillation noise by the interface circuit in Fig. 3.2 is analyzed: V_{CAR} is the output signal of the carrier generator at a frequency of several to tens of MHz. The signal generated by the motion of the detection frame (kHz) is modulated by the carrier, so the signal frequency through which the parasitic resistor passes becomes the carrier frequency, which is three orders of magnitude higher than the kHz frequency of the carrier-free interface analyzed earlier. Since the scintillation noise is inversely proportional to the frequency of the current passing through the target resistor [116], the equivalent voltage of the scintillation noise generated by the parasitic resistor, V_{FNps}, is reduced by more than 30 times.

Second, the interface circuit in Fig. 3.2 also has good suppression of common-mode noise: when the positive half period of the carrier wave arrives, diodes D_1 and D_3 turn on, and the voltages V_{L+} and V_{L-} are connected to the positive and negative inputs of the instrument amplifier through a diode drop. It is not difficult to obtain the following equation about noise:

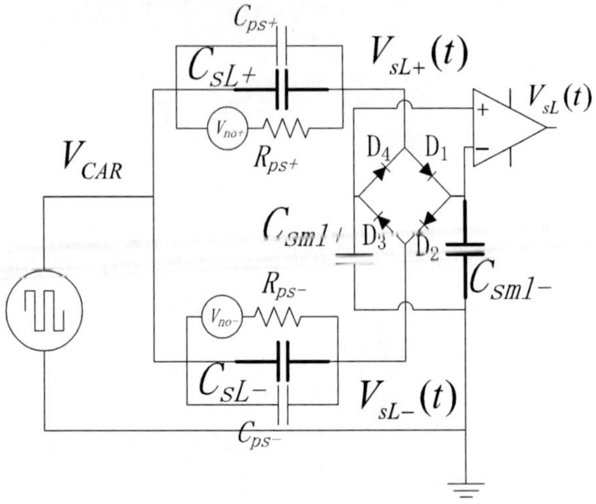

Fig. 3.2 The output interface circuit

$$V_{sLnoise} = K_{preamp}\left[(V_{noiseL-} - V_{D3}) - (V_{noiseL+} - V_{D1})\right] \tag{3.1}$$

$$\begin{cases} V_{noiseL+} = V_{CAR} - V_{ELENps+} - V_{FNps+} - i_{ps+}R_{ps+} \\ V_{noiseL-} = V_{CAR} - V_{ELENps-} - V_{FNps-} - i_{ps-}R_{ps-} \end{cases} \tag{3.2}$$

where, $V_{sLnoise}$ is the noise component output by the interface; K_{preamp} is the magnification factor of the instrument amplifier; $V_{noiseL+}$ and $V_{noiseL-}$ are the noise components in V_{sL+} and V_{sL-} respectively; V_{D3} and V_{D1} are the voltage drops of diode D_3 and D_1; I_{ps+} and I_{pS-} are the currents flowing through the parasitic resistor, respectively. According to the correlation expression of electronic thermal noise [107–115], it can be obtained:

$$\begin{aligned} V_{sLnoise} = K_{preamp}\Bigg[\Bigg(&V_{CAR} - \sqrt{4k_BTR_{ps-}B_{noise}} - \sqrt{K_{FN}R_{ps-}^2\frac{i_{ps-}}{f_{cur}}B_{noise}} - i_{ps-}R_{ps-} - V_{D3}\Bigg) \\ &-\Bigg(V_{CAR} - \sqrt{4k_BTR_{ps+}B_{noise}} - \sqrt{K_{FN}R_{ps+}^2\frac{i_{ps+}}{f_{cur}}B_{noise}} - i_{ps+}R_{ps+} - V_{D1}\Bigg)\Bigg] \end{aligned} \tag{3.3}$$

where, f_{cur} is the frequency of the current flowing through the parasitic resistor. Due to the symmetrical structure and packaging, the parasitic resistance and capacitance R_{ps+} and R_{ps-}, C_{ps+} and C_{ps-} can be approximately equal, and at the same time, $V_{D3} = V_{D1}$, the above equation can be simplified as:

$$V_{sLnoise} = K_{preamp}\left(\sqrt{K_{FN}R_{ps+}^2\frac{i_{ps+}}{f_{cur}}B_{noise}} + i_{ps+}R_{ps+} - \sqrt{K_{FN}R_{ps-}^2\frac{i_{ps-}}{f_{cur}}B_{noise}} - i_{ps-}R_{ps-}\right) \tag{3.4}$$

In the static case, it can be considered that the detection capacitors C_{sL+} and C_{sL-}, and the precision capacitors C_{sml+} and C_{sml-} are approximately equal respectively, so the circuits passing through the two loops of V_{CAR-}, V_{sL+}, -GND and V_{CAR-}, V_{sL-},

-GND are almost identical, so the current on the parasitic resistance is almost equal. Then the right side of Eq. (3.4) is nearly zero. In the dynamic case, when the difference between the two detection capacitors becomes larger, there is an obvious difference between the two loops. Then, $V_{sLnoise}$ in the above formula will have residual noise component, but its size will also be greatly weakened by this difference form.

3.2.3 Ring Diode Interface Circuit Working Principle

In the previous section, the principle of noise suppression by differential interface circuit is introduced. In this section, the parasitic elements in the circuit in Fig. 3.2 are ignored for the convenience of description, and the ring diode detection circuit is shown in Fig. 3.3 [118, 119]:

When the ring diode circuit is charged, there is positive half-cycle step input of square wave with amplitude of V_{CAR}, then diode D_3 and D_1 are turned on, and the circuit charges the precise capacitors C_{sml+} and C_{sml-}. According to the circuit switching theorem, equations of transient process of the circuit can be established:

$$i_{L+}(t) = C_{sL+}\frac{dU_{CsL+}(t)}{dt} = C_{sml-}\frac{dU_{Csml-}(t)}{dt} \tag{3.5}$$

$$R_{CAR}i_{L+}(t) + U_{CsL+}(t) + U_{Csml-}(t) + V_{D1} = V_{CAR} \tag{3.6}$$

$$U_{CsL+}(0) = U_{Csml-}(0) = 0 \tag{3.7}$$

where i_{L+} is the current flowing through D_1; U_{Cs+} is C_{s+} voltage at both ends; U_{Csml-} is the voltage at both ends of C_{sml-}; $R_{CAR} = 100\ \Omega$ is the carrier generator internal

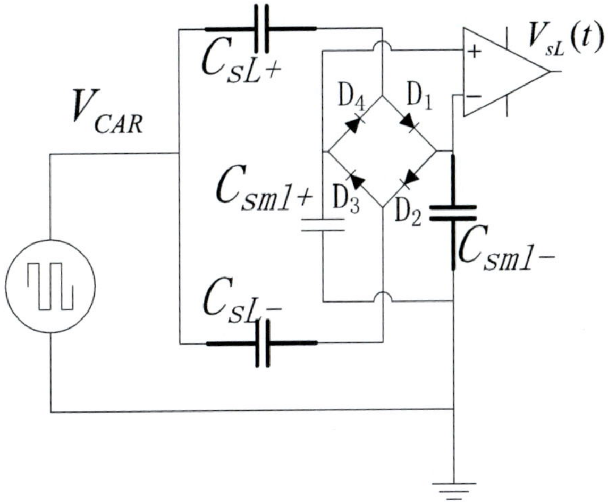

Fig. 3.3 Ring diode detection interface circuit

resistance. The above differential equation can be obtained:

$$U_{Csml-}(t) = \frac{C_{sL+}(V_{CAR} - V_{D1})}{C_{sL+} + C_{sml-}}\left\{1 - \exp\left[-\frac{t(C_{sL+} + C_{sml-})}{R_{CAR}C_{sL+}C_{sml-}}\right]\right\} \tag{3.8}$$

The voltage changes of carrier excitation from $+V_{CAR}$ to $-V_{CAR}$ can be analyzed by using the same method. Since the detection capacitance is all at pF level, the precision capacitance is 200 pF, and the carrier frequency is several MHZ, the time constant of the above equation should be far less than the half-cycle of carrier excitation, and the charging process can be ignored. After ignoring the diode pressure drop, the voltage of the above equation and diode D_3 side can be expressed as:

$$\begin{cases} U_{Csml-}(t) = \dfrac{C_{sL+}V_{CAR}}{C_{sL+} + C_{sml-}} \\ U_{Csml+}(t) = \dfrac{C_{sL-}V_{CAR}}{C_{sL-} + C_{sml+}} \end{cases} \tag{3.9}$$

When there is angular rate input, the detection capacitance changes, let $C_{sL+} = C_{sL0} + \Delta C_{sL}$, $C_{sL-} = C_{SL0} - \Delta C_{sL}$, $C_{sml+} = C_{sml-} = C_{sml}$, so the output voltage of the pre-stage amplifier can be expressed as:

$$V_{sL} = K_{preamp}(U_{Csml+} - U_{Csml-}) = 2K_{pre}V_{CAR}\frac{-C_{sml}\Delta C_{sL}}{(C_{sL0} + C_{sml})^2 - \Delta C_{sL}^2} \tag{3.10}$$

Have $C_{sml} + C_{sL0} >> \Delta C_{sL}$, then the above formula can be further simplified as:

$$V_{sL} \approx 2K_{preamp}V_{CAR}\frac{-C_{sml}\Delta C_{sL}}{(C_{sL0} + C_{sml})^2} \tag{3.11}$$

As can be seen from the above equation, under the condition that the capacitance change is not very large, although the output voltage of the front discharge interface is directly proportional to the capacitance change, its nonlinearity gradually decreases with the increase of ΔC_{sL}. Combined with the analytical formula (2.40) for the variation of capacitor of synovial comb tooth in Chap. 2, V_{sL} has a good linear relationship with the displacement of frame when the detection displacement is very small. In addition, when the structure is determined, increasing the carrier amplitude V_{CAR} and decreasing the precision capacitance C_{sml} can increase the sensitivity of output signal to the detection displacement. When detecting displacement is larger (angular velocity larger input state), the interface will deteriorate gradually, and the linear relationship between the factors influence the scale factor nonlinearity and asymmetry degree and other related indicators, this problem can be by examining the closed-loop optimization of force feedback (inspection when closed loop displacement is almost zero), Chap. 6 in the gyro the working state of experimental data can be very good certify that the above analysis.

3.3 Dual-Mass Linear Vibration Silicon-Based MEMS Gyroscope Structure System Model

In order to further study the characteristics of dual-mass linear vibration silicon-based MEMS gyroscope structure, this section establishes a system model of ideal gyro structure based on the motion equation of gyro studied in Chap. 2, and adds the equivalent interference force of mechanical thermal noise proposed in the previous two sections [120, 121]. In addition, the following chapters will add the influencing factors of machining error and static compensation and feedback modules existing in actual engineering on the basis of the structural ideal system model in this section, providing a simulation basis for the subsequent work. There needs to be pointed out that the book of gyroscope structure is under the condition of the vacuum working (approximately thousands of quality factor), and when the input angular rate is larger (for example, when the angular velocity gyroscope full range input), driving mode also will be affected by the Coriolis acceleration effect sensing modal movement, so the item $2m_c\Omega_z\dot{y}$ should join in the type (2.22), After adding mechanical thermal noise and detection, the combined Eq. (2.22) is as follows:

$$\begin{cases} m_x\ddot{x} + c_x\dot{x} + k_xx + F_{MTNx} = F_{dx} + 2m_c\Omega_z\dot{y} \\ m_y\ddot{y} + c_y\dot{y} + k_yy + F_{MTNy} = -2m_c\Omega_z\dot{x} \end{cases} \tag{3.12}$$

According to the expression content in Formula (3.23), an equivalent system model of gyro structure containing mechanical thermal noise is established in Simulink, as shown in Fig. 3.4 The dark part is the equivalent model of driving mode, and the light part is the detection mode. In the whole system, there are two inputs: driving voltage and input angular velocity. The three outputs are driving displacement, Coriolis force (easy to observe) and sensing displacement. Based on the structural data in Chap. 2, you can calculate the corresponding parameter values in Fig. 3.4, as shown in Table 3.1.

3.4 Silicon Based MEMS Gyroscope Structure Drive Technology

It can be seen from the gyroscopic driving mode motion equation in Chap. 2 that the silicon micromechanical gyroscope works on the premise of driving mode constant amplitude vibration, so it is necessary to design a driving closed-loop to meet the above conditions. At the same time, in order to maximize the displacement of the driving mode, the driving frequency is usually equal to the resonant frequency of the driving mode. At this point, the phase Angle of the driving displacement is set as $\varphi_x = 90°$, then the phase difference between the driving force and the driving displacement

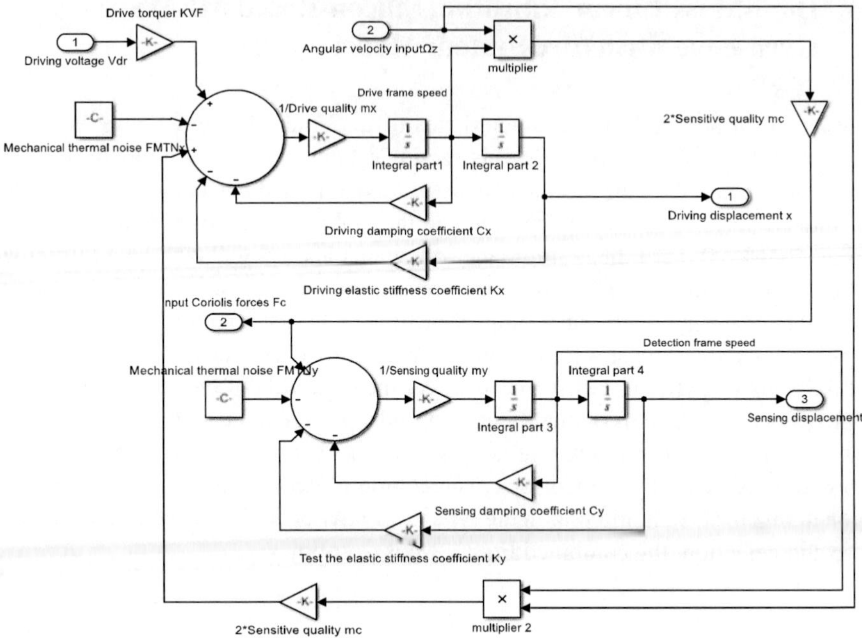

Fig. 3.4 Equivalent system model of gyro structure

Table 3.1 Simulation parameters of gyro structure system model

Parameter name	Parameter values
Driving mass m_x	1.2355×10^{-6} kg
Sensing mass $m_y(m_c)$	1.1232×10^{-6} kg
Driving stiffness coefficient k_x	593.9 N/m
Sensing stiffness coefficient k_y	532.0 N/m
Driving damping coefficient c_x	9.01×10^{-6} N/m/s
Sensing damping coefficient c_y	9.43×10^{-6} N/m/s
Driving thermal noise interference force F_{MTNx}	8.89×10^{-13} N
Sensing thermal noise interference force F_{MTNy}	9.05×10^{-13} N

is 90°. Therefore, there are two methods in the design of gyro driving loop: phase-locked loop (PLL) method and self-excited oscillation control method. But due to the phase angle of driving closed-loop PLL and gain control are mutually coupled, so this method is not conducive to optimize the phase and amplitude control, combined with the phase-locked loop structure is not suitable for control quality factor higher gyroscope (and gyro structure usually adopts vacuum encapsulation led drive mode of high-quality factor) [70]. Therefore, this book chooses to design the gyroscope driving loop by self-excited oscillation mode. The controller works in the mode of automatic gain, and the phase shift of 90° in the loop is realized through the phase shift link.

3.4.1 Silicon Based MEMS Gyroscope Self-Excited Drive Closed-Loop Scheme

The premise of the self-excited oscillation of the driving loop is to satisfy the gain of the whole closed-loop A_{dclose} >1 and the phase angle of the closed-loop $\varphi_{dclose} = 2k\pi$, where k is an integer. The larger A_{dclose} is, the more sensitive the dynamic response of the system is and the shorter the vibration start time is. According to the push–pull electrostatic driving method in Chap. 2, the driving force is determined by the product of the amplitude of the DC component and the AC component of the driving voltage. Therefore, in order to simplify the control algorithm and improve the control accuracy, the book adopts the method of fixed drive direct flow and changing the amplitude of AC voltage through AGC technology to achieve the purpose of controlling the electrostatic driving force (because the DC voltage can be directly provided by the voltage stabilizer chip, with good stability at room temperature and full temperature).The equivalent transfer function of the driving mode can be equivalent to a bandpass filter with a very high quality factor, so only the center frequency ω_x and its adjacent signals are retained in the loop. Therefore, this method can ensure that the gyroscope driving mode works at its resonant frequency with a constant amplitude [70]. At this time, the maximum vibration amplitude can be obtained by driving displacement, which is beneficial to the detection and extraction of Coriolis signal.

Figure 3.5 shows the self-excited oscillation driving closed-loop. The system takes the gyroscope structure driving mode displacement as the control object. Among them, inside the dotted box is the equivalent model of the driving mode of the gyro structure (Fig. 3.4 does not include the X/C converter, and this module is placed outside the gyro head model to observe x).The driving displacement signal inside the gyro structure is converted into the change of driving detection capacitance ΔC_{dc} by the synovial comb tooth, and then is further converted and amplified by the interface circuit to obtain V_x (the differential output of the left and right mass block driving detection comb tooth), which is divided into two parts after 90° precision phase shifting circuit: One part serves as the modulation reference V_{dac} that drives the AC component, and the other part of the amplitude signal V_{dacA} is extracted by the full-wave rectifier and low-pass filter. After superposition of signal amplitude V_{dacA} and reference voltage reference V_{ref}, the integrator forms control signal V_{dI}, which becomes AC component V_{ac} of driving voltage after modulation-by-modulation reference V_{dac}. The DC component of the driving voltage, V_{dc} and V_{ac}, becomes the driving voltage signal, V_{dr}. After passing through the torque converter K_{VF}, the driving electrostatic force is formed to act on the equivalent model of the

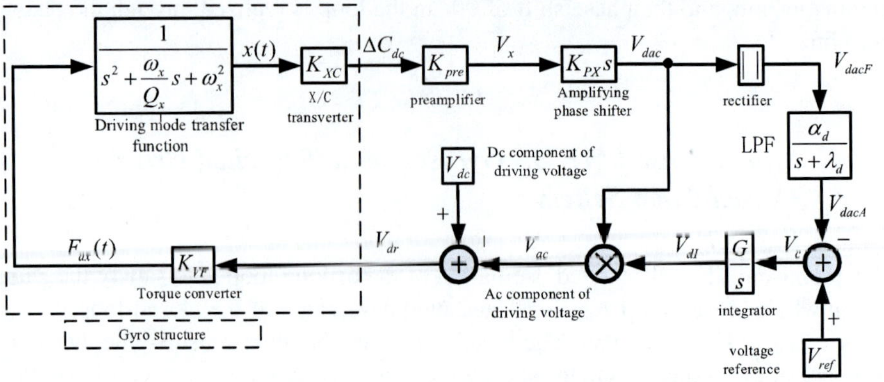

Fig. 3.5 Block diagram of self-excited drive closed-loop

gyroscope's driving mode. The whole system has simple principle, convenient circuit realization and easy parameter debugging.

3.4.2 *Silicon Based MEMS Gyroscope Drive Loop Working Point Analysis*

This section uses "average method" to study the stability of driving loop. According to the analysis of the electrostatic force of the comb tooth in Chap. 2, the push–pull electrostatic driving force of the driving mode can be expressed as [70, 79, 122]:

$$F_{dx} = 2\frac{\partial C_d}{\partial x} V_{dc} V_{ac} \tag{3.13}$$

where, C_d is the driving capacitance of single side comb tooth. Substitute the above equation into the driving modal motion equation of Eq. (2.23), which can be obtained by combining with Fig. 3.8

$$\ddot{x} + \frac{\omega_x}{Q_x}\dot{x} + \omega_x^2 x = 2\frac{\partial C_d}{m_x \partial x} V_{dc} V_{dI}(t) K_{XC} K_{pre} K_{PX} \dot{x} \tag{3.14}$$

$$\dot{V}_{dI}(t) = G(V_{ref} - V_{dacA}) \tag{3.15}$$

$$\dot{V}_{dacA} = \left|K_{XC} K_{pre} K_{PX} \dot{x}\right| \alpha_d - \lambda_d V_{dacA}(t) \tag{3.16}$$

Suppose the driving displacement is:

$$x(t) = a_x(t)\cos(\omega_x t + \varphi_d(t)) \tag{3.17}$$

where $a_{x(t)}$ and $\varphi_{d(t)}$ are the motion amplitude and phase of the driving mode respectively, then the motion speed of the driving mode can be expressed as:

$$\dot{x} = \dot{a}_x \cos(\omega_x t + \varphi_d(t)) - a_x(t)\sin(\omega_x t + \varphi_d(t))(\omega_x + \dot{\varphi}_d) \tag{3.18}$$

According to the "average method" there are:

$$\dot{a}_x \cos(\omega_x t + \varphi_d(t)) - a_x(t)\dot{\varphi}_d \sin(\omega_x t + \varphi_d(t)) \equiv 0 \tag{3.19}$$

Then formula (3.18) can be simplified as:

$$\dot{x} = -a_x(t)\omega_x \sin(\omega_x t + \varphi_d(t)) \tag{3.20}$$

In addition, the motion acceleration of the driving mode is:

$$\ddot{x} = -\dot{a}_x\omega_x \sin(\omega_x t + \varphi_d(t)) - a_x(t)\omega_x \cos(\omega_x t + \varphi_d(t))(\omega_x + \dot{\varphi}_d) \tag{3.21}$$

Substituting Eqs. (3.17), (3.20) and (3.21) into Eq. (3.14), we can get:

$$\begin{aligned} &-(\dot{a}_x\omega_x + a_x\frac{\omega_x^2}{Q_x})\sin(\omega_x t + \varphi_d) - a_x\dot{\varphi}_d\omega_x\cos(\omega_x t + \varphi_d) \\ &= -2\frac{\partial C_d}{m_x \partial x}V_{dc}V_{dI}(t)K_{XC}K_{pre}K_{PX}a_x\omega_x \sin(\omega_x t + \varphi_d) \end{aligned} \tag{3.22}$$

Substituting Eq. (3.19) into the above equation, it can be obtained:

$$\dot{a}_x = -a_x\frac{\omega_x}{Q_x}\sin^2(\omega_x t + \varphi_d) - 2\frac{\partial C_d}{m_x \partial x}V_{dc}V_{dI}(t)K_{XC}K_{pre}K_{PX}a_x \sin^2(\omega_x t + \varphi_d) \tag{3.23}$$

$$\dot{\varphi}_d = -\frac{\omega_x}{2Q_x}\sin(2\omega_x t + 2\varphi_d) - \frac{\partial C_d}{m_x \partial x}V_{dc}V_{dI}(t)K_{XC}K_{pre}K_{PX}\sin(2\omega_x t + 2\varphi_d) \tag{3.24}$$

Substituting Eq. (3.20) into Eq. (3.16) and based on the "average method" theory, the mean values of Eqs. (3.15), (3.16), (3.23) and (3.24) within one cycle ($T = 2\pi/\omega_x$) can be obtained:

$$\dot{V}_{dI}(t) = \frac{1}{T}\int_0^T G(V_{ref} - V_{dacA})dt \tag{3.25}$$

$$\dot{\overline{V}}_{dacA} = \frac{1}{T}\int_0^T \left|K_{XC}K_{pre}K_{PX}a_x(t)\omega_x \sin(\omega_x t + \varphi_d(t))\right|\alpha_d - \lambda_d V_{dacA}(t)dt \tag{3.26}$$

$$\dot{\overline{a}}_x = \frac{1}{T}\int_0^T -a_x(\frac{\omega_x}{Q_x} + 2\frac{\partial C_d}{m_x \partial x} V_{dc} V_{dI}(t) K_{XC} K_{pre} K_{PX}) \sin^2(\omega_x t + \varphi_d) dt \tag{3.27}$$

$$\dot{\overline{\varphi}}_d = \frac{1}{T}\int_0^T -(\frac{\omega_x}{2Q_x} + \frac{\partial C_d}{m_x \partial x} V_{dc} V_{dI}(t) K_{XC} K_{pre} K_{PX}) \sin(2\omega_x t + 2\varphi_d) dt \tag{3.28}$$

Then

$$\dot{\overline{V}}_{dI}(t) = G(V_{ref} - \overline{V}_{dacA}) \tag{3.29}$$

$$\dot{\overline{V}}_{dacA} = \frac{2}{\pi} \overline{a}_x \omega_x K_{PX} \alpha_d \left| K_{XC} K_{pre} \right| - \lambda_d \overline{V}_{dacA} \tag{3.30}$$

$$\dot{\overline{a}} = -\frac{\overline{a}}{2}(\frac{\omega_x}{Q_n} + 2\frac{\partial C_d}{m_x \partial x} V_{dc} K_{XC} K_{pre} K_{PX} \overline{V}_{dI}) \tag{3.31}$$

$$\dot{\overline{\varphi}}_d = 0 \tag{3.32}$$

Let the right end of formula (3.29), (3.30) and (3.31) be zero, then:

$$\overline{V}_{dcaA0} = V_{ref} \tag{3.33}$$

$$\overline{a}_{x0} = \frac{\pi \lambda_d V_{ref}}{2\omega_x K_{PX} \alpha_d \left| K_{XC} K_{pre} \right|} \tag{3.34}$$

$$\overline{V}_{dI0} = -\frac{m_x \partial x \omega_x}{2\partial C_d Q_x V_{dc} K_{XC} K_{pre} K_{PX}} \tag{3.35}$$

It can be seen from Eq. (3.34) that the driving mode vibration amplitude has only one stable operating point, at which the amplitude is only related to the reference voltage, the driving mode resonant frequency and the driving mode closed-loop gain.

3.4.3 Dual-Mass Linear Vibration Silicon-Based MEMS Gyroscope Drive Loop System Simulation

In the second chapter, it is pointed out that the design value of the top–bottom distance of the synovium comb tooth structure is 20 μm, and a large vibration amplitude is needed to improve the mechanical sensitivity of the structure, but too large displacement will lead to the deterioration of the nonlinear relationship between the detection capacitance and displacement. So the driving mode selected in the actual

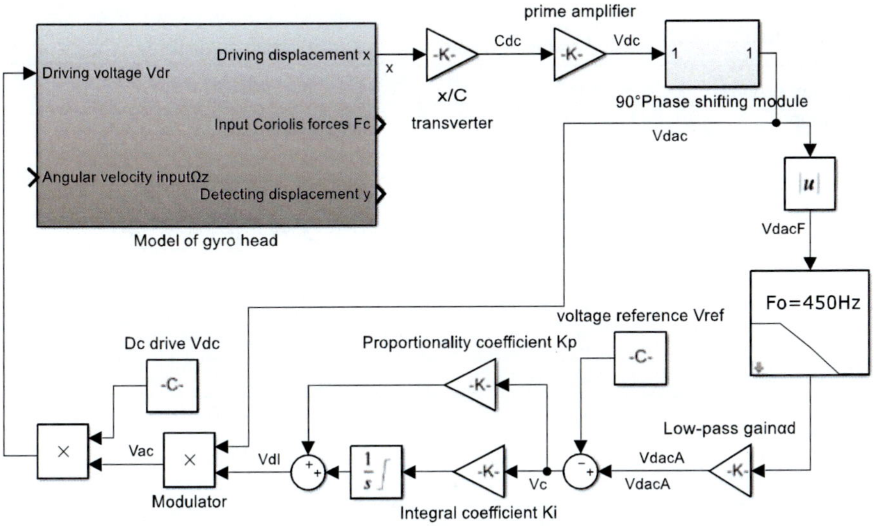

Fig. 3.6 Simulation block diagram of gyro-driven mode self-excited closed-loop drive

project motion amplitude is 1.5 μm, on the one hand to ensure larger driving displacement and mechanical sensitivity, help Coriolis signal detection in order to achieve a high SNR, on the other hand also to overshoot and other factors in the process of start retains sufficient margin, at the same time, also guarantees the linearity of the drive displacement detection. Based on this working point, the parameters of the controller are optimized to balance the starting time and overshoot and avoid the comb tooth collision. Figure 3.6 shows the simulation block diagram of gyro-driven mode self-excited closed-loop drive, in which the "gyro-head model" module is shown in Fig. 3.4.

In accordance with the design in Fig. 3.5, a proportional link is added into the integral link to reduce the saturation of the integrator caused by the input misalignment error of the operational amplifier. The driving voltage is directly expressed as the product of the DC and AC components in the figure above. The 90° phase shift module is a precise phase shift for the resonant frequency of the driving mode, which is realized by the method of one-quarter period delay. Oscilloscopes observe the displacement signals that drive the modes. Table 3.5 shows the simulation parameters of the above system. Substitute the parameters in the table into Fig. 3.6 to obtain the system simulation figure as shown in Fig. 3.7.

Figure 3.7 shows the simulation results of driving displacement x. It can be seen that the maximum amount of x in the starting process is about 3 μm, and it stabilates near 1.5 μm within 1 s, which meets the requirements of fast and stable. Moreover, its vibration frequency is the same as the resonant frequency of the driving mode, which proves the correctness of the self-excited driving closed-loop model. Due to the large driving force, the simulation results also prove that the influence of mechanical thermal noise interference signal on the driving mode can be ignored.

Table 3.5 Simulation parameters of gyroscopic self-excited driving closed-loop

Parameter name	Parameter values
K_{XC}	8.5×10^{-7}F/m
K_{pre}	-8.374×10^{11} V/F
K_{PX}	1
α_d	10
λ_d	3.54×10^{-3} s
V_{ref}	5 V
K_p	0.6
K_i	65
V_{dc}	5 V
K_{VF}	5.777×10^{-7} N/V

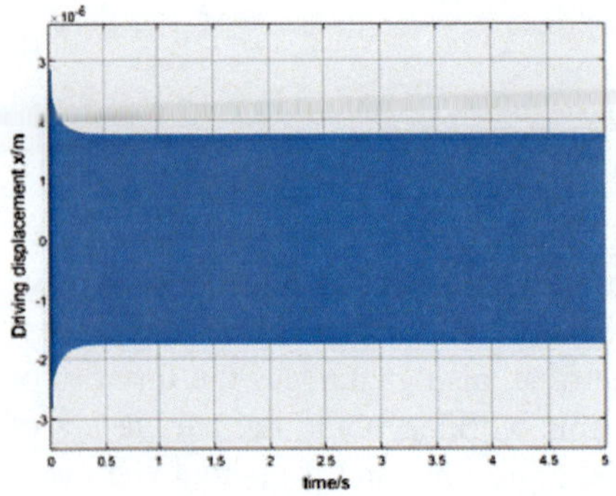

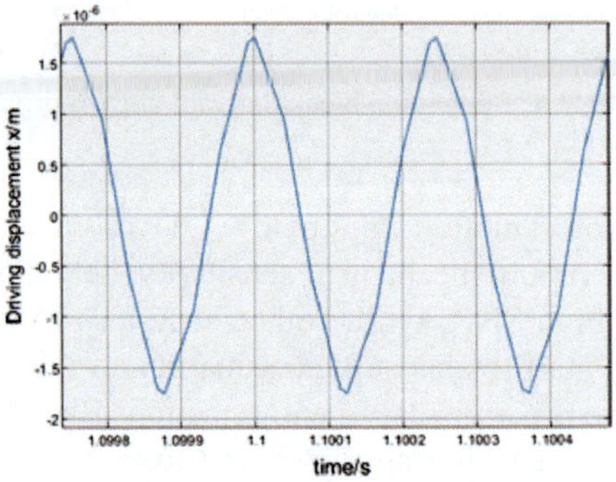

Fig. 3.7 Simulation diagram of driving displacement x in self-excited driving closed-loop (left) and local enlargement diagram (right)

At the same time, the 90° phase shift link, rectification and filtering link, integrator input, integrator output control, driving voltage AC and driving voltage DC should also be monitored in the self-excitation driving closed-loop. These observations are shown in Figs. 3.8, 3.9, 3.10, 3.11, 3.12, and 3.13.

It can be seen from the six figures that the 90° phase shift link can accurately phase shift the input signal (solid line signal) for a quarter of a cycle, and the amplitude and frequency remain unchanged. The AC amplitude V_{dacA}, which is related to the driving displacement measurement x, remains at about 5 V after stabilization, which is the same as the reference voltage Vref. The integrator input Vc is the difference between the reference voltage V_{ref} and V_{dacA}, which is stable at about 0 V. The output value of the integrator is in a stable state after the system is stabilized within 1 s. The AC component of the driving voltage, Vac, has been well maintained at about 0.32 V peak-to-peak value after stabilization; The driving force is proportional to the product of ac and DC components of the driving voltage. In the simulation results, the product is also stable at about $1.6V^2$ peak-to-peak value. It can be seen from the above simulation results that each node in the self-excited driving closed-loop

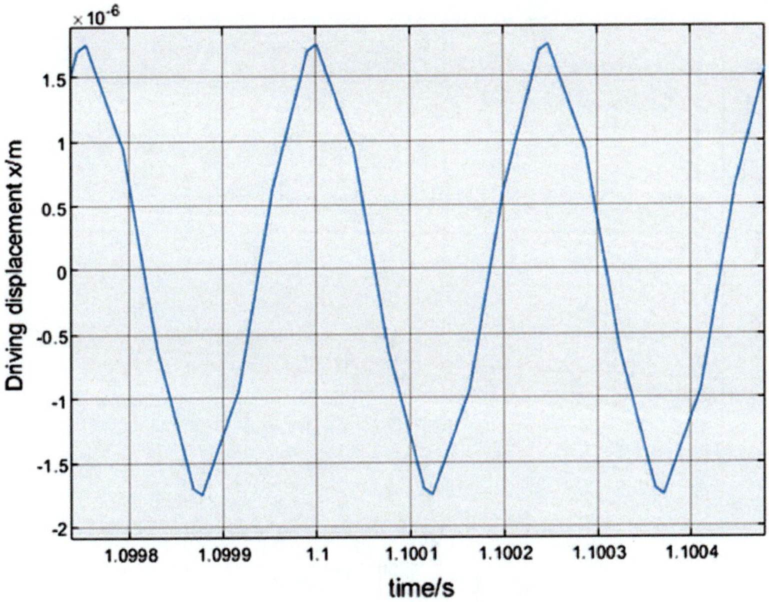

Fig. 3.8 Input and output simulation results of a 90° phase shifter

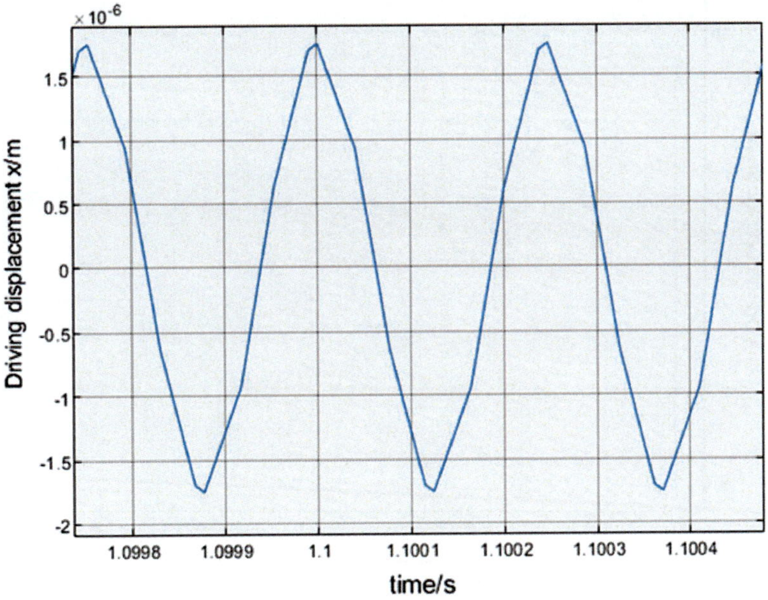

Fig. 3.9 V_{DACA} output results simulation results

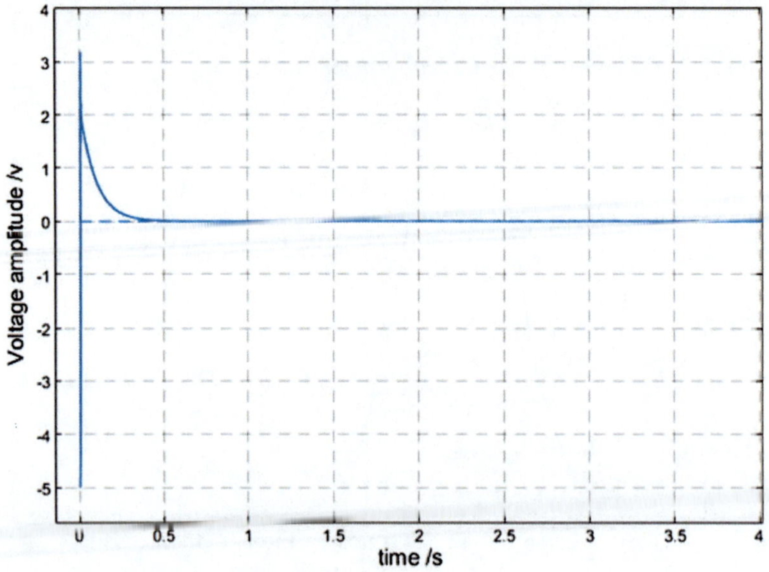

Fig. 3.10 Vc simulation results of the integrator input

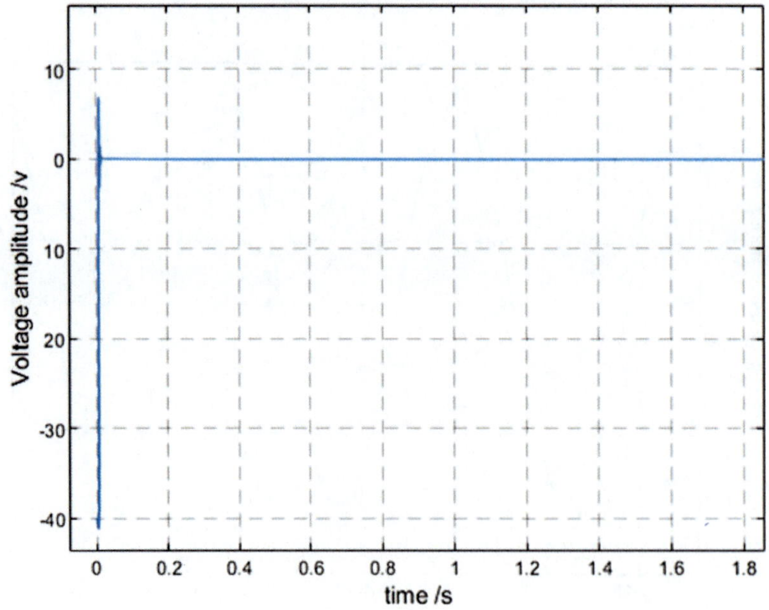

Fig. 3.11 V_{dI} simulation results of the integrator output

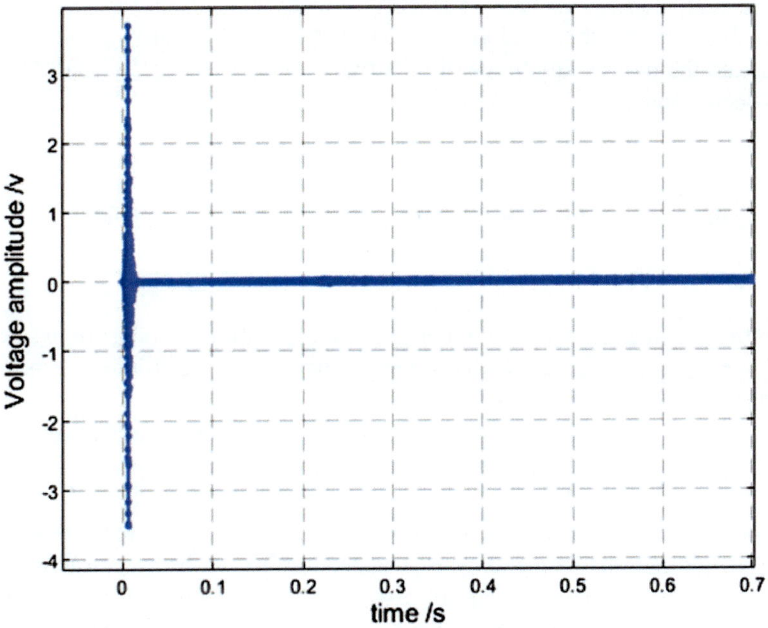

Fig. 3.12 Simulation results of AC component of driving voltage

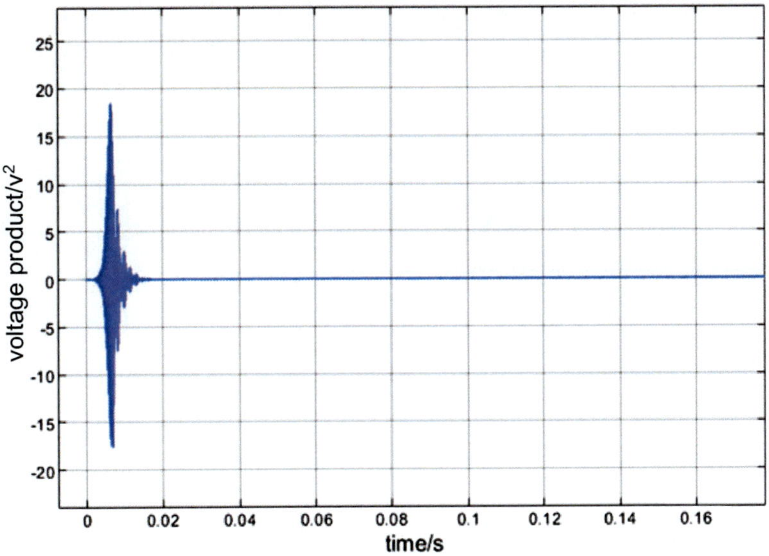

Fig. 3.13 Simulation results of AC-DC driving voltage product

is in a stable working state, and the resonant frequency of the driving mode is well tracked, and the motion amplitude of the driving mode is well stabilized at the set value selected above.

3.4.4 Dual-Mass Linear Vibration Silicon-Based MEMS Gyroscope Drive Loop Test

In order to verify the actual working state and stability of the preceding driving closed-loop system, this section conducts a full-temperature test on the driving loop based on the actual circuit, and the results are shown in Fig. 3.14. During the experiment, the temperature was firstly raised to 60°C, and then the temperature was lowered at the maximum cooling rate of the temperature control box for one hour, and the Vx data was recorded at the sampling rate of 1 Hz. After dropping to −40°C, the temperature was kept at −40°C for one hour and the inside and outside of the gyroscope was ensured. After that, the temperature was heated to 60°C at the highest temperature return rate of the temperature control box, and the temperature was kept for one hour to ensure the same temperature inside and outside the gyro. The above heating process samples the entire gyro output.

As the driving displacement cannot be measured directly, it can be seen from Fig. 3.5 that the driving displacement is in a linear relationship with the forward discharge output Vx, so its value can indirectly reflect the driving displacement. According to the data in Table 3.5, Vx should be about 1.1 V when the driving displacement is 1.5 μm. The value shown in Fig. 3.14 is consistent with the theoretical calculation value. The driving displacement varies from 1.57 μm to 1.56 μm in

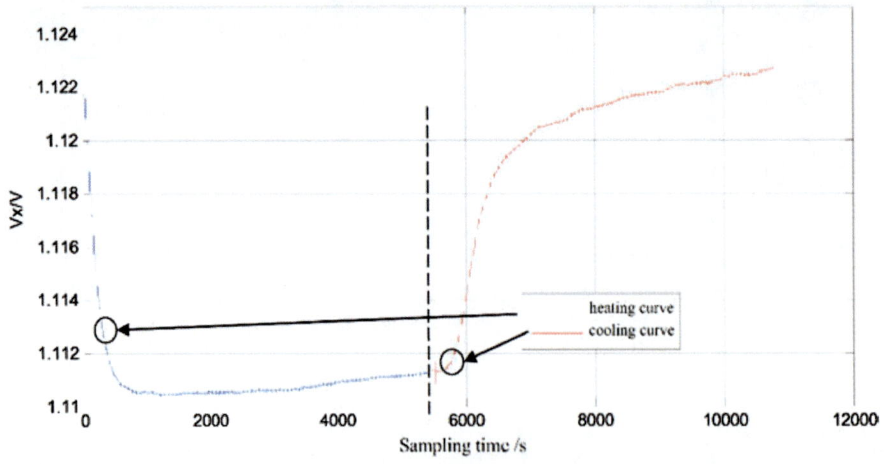

Fig. 3.14 Driving displacement full temperature experiment

the full temperature range, which basically meets the requirement of constant amplitude and stability at full temperature. However, with the improvement of gyroscope accuracy, the requirement of stable amplitude at full temperature drive is becoming more and more stringent. Temperature compensation and digital circuit technology can further improve the characteristics in this area.

3.5 Silicon Based MEMS Gyroscope Structural Noise

3.5.1 Structural Noise Analysis

The silicon micromechanical gyroscope structure is mainly composed of silicon, glass and metal leads, and its noise mainly consists of the following parts [160].

1. Mechanical Thermal Noise

The noise is widely existing in micro mechanical gyro structure, mainly caused by molecular movement within the silicon structure, usually think mechanical thermal noise as the limit of the mechanical structure sensitivity, mechanical thermal noise can be made of Gaussian noise in the form of expression, at the same time, the mechanical thermal noise can be equivalent to parallel a damper in the structure, mechanical damping force produced by thermal noise can expression is:

$$F_{MTN} = \sqrt{4k_B T c B} \tag{3.36}$$

where $k_B = 1.38 \times 10^{-23}$ J/K is Boltzmann constant; T is temperature; C is the damping coefficient; B is the noise bandwidth.

2. Electronic Thermal Noise

The noise, also known as Johnson noise, is generated by the electronic thermal motion noise of the equivalent resistance in the structure, and the voltage generated by the noise is expressed as:

$$V_{ETN} = \sqrt{4k_B T R B} \tag{3.37}$$

where, R is the resistance value of electronic thermal noise. The distributed resistors R_{pdc+}, R_{pdc-}, R_{ps+} and R_{ps-} have large resistance values, which should be the main source of electronic thermal noise.

3. Scintillation Noise (Flicker Noise)

The scintillation noise, also known as 1/F noise, is generated by the fluctuation of the conductivity of the semiconductor device, and the equivalent voltage generated by it is expressed by the following formula:

$$V_{FN} = \sqrt{K R^2 \frac{I}{f} B} \quad (3.38)$$

where, K is a constant, determined by the material type and its geometric structure; I am the current flowing through the resistance; F is the frequency of current I.

4. Constant Frequency Noise of Driving Signal

This part of the signal is described in detail in the next section.

3.5.2 Structural Noise Test

In order to better reflect the noise characteristics of the structure, a high-speed digital acquisition module (sampling rate 64 kHz) is adopted to collect and analyze the output signal of the gyro structure under the normal operation of the drive mode. Since the left and right mass blocks are designed symmetrically, the book takes one of the mass blocks as an example to test the relevant noise: By grounding the detection circuit V_{SL-} as shown in Fig. 3.2, the interface circuit becomes the unilateral sensing

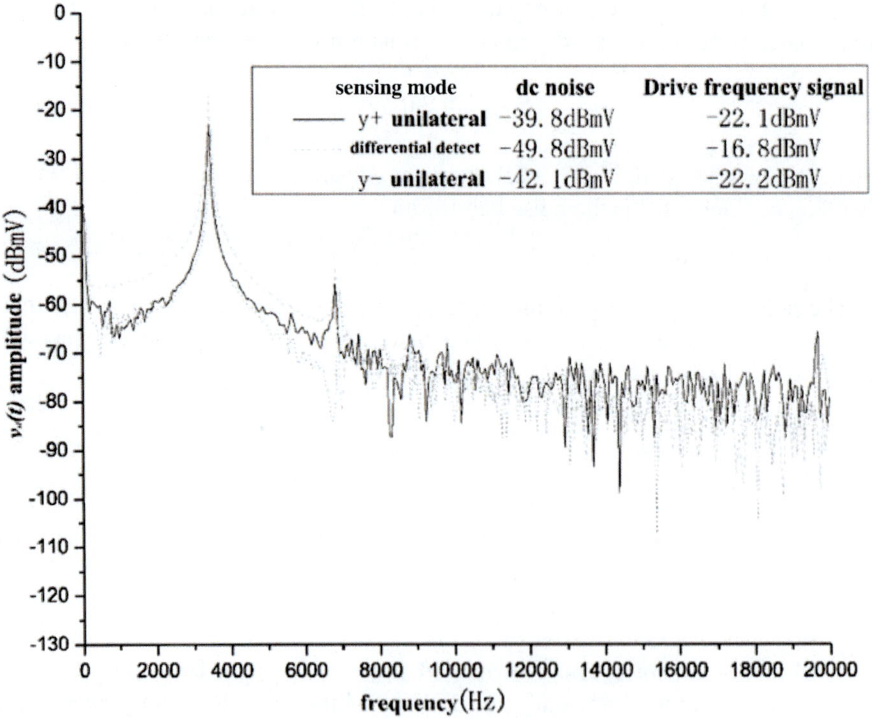

Fig. 3.15 Interface detection test curve

mode and collects V_{sL} signals. Similarly, V_{sL+} grounding can obtain the unilateral detection mode on the other side. Finally, the differential detection mode is adopted and the following spectrum diagram can be obtained after Fourier transform. The amplitude of driving frequency point signal increases (Fig. 3.15).

According to Formula 3.38, without changing other parameters, the scintillation noise is related to the frequency of current passing through the resistance, and the higher the carrier frequency is, the lower the noise is. Therefore, the scintillation noise in the structure can be tested by changing the carrier frequency, and the carrier frequency is set at 0.5 MHz, 1 MHz, 5 MHz and 10 MHz respectively, and differential detection method is adopted. Spectrum analysis was conducted on the data acquisition results. The curve is shown in Fig. 3.16. With the increase of carrier frequency, the DC noise component decreases and the amplitude of driving frequency signal basically remains unchanged, which is consistent with the theoretical analysis results.

It can be seen from Eqs. 3.36 and 3.37 that structural mechanical thermal noise and electronic thermal noise are related to temperature. In order to verify the influence of these two noises on the structure, the amplitude of the two noises can be changed by changing the temperature of the structure environment, and the output signal of the structure can be measured, as shown in Fig. 3.17. It can be seen from the figure that as the temperature rises, the DC noise increases and the amplitude of the driving frequency signal basically remains unchanged, which is consistent with the results

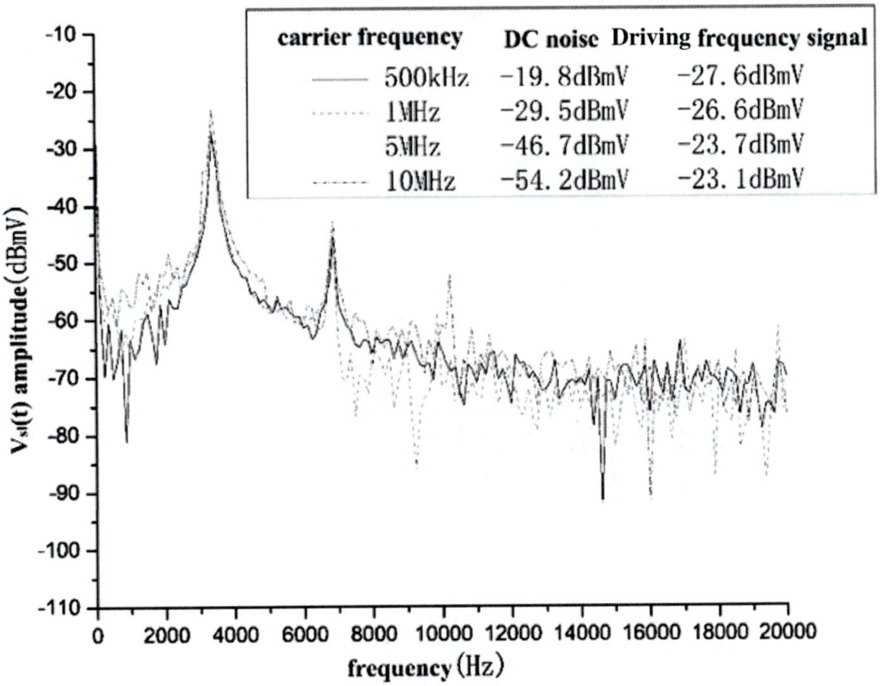

Fig. 3.16 Scintillation noise test curve

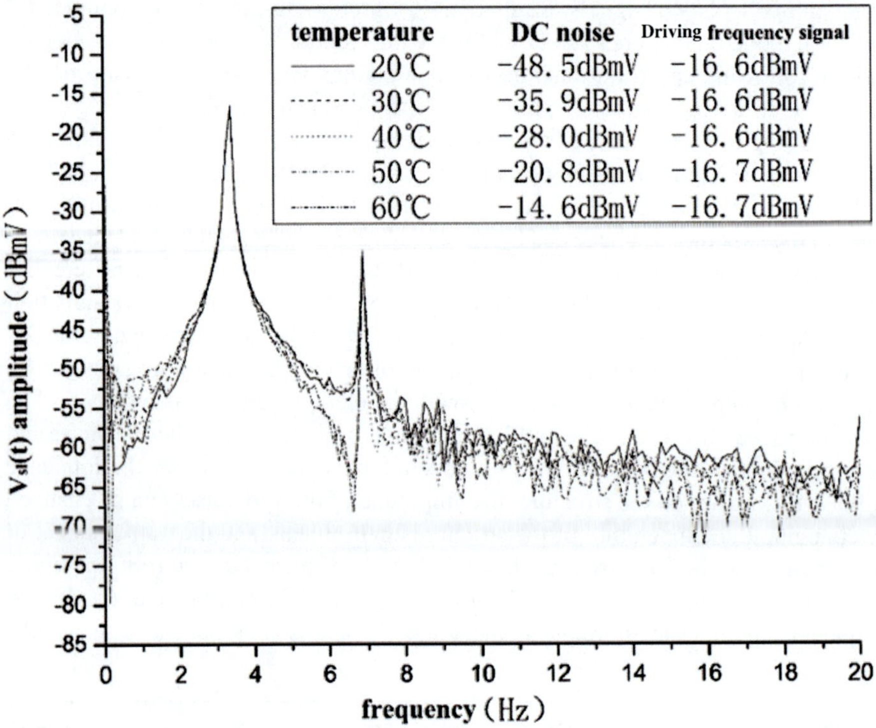

Fig. 3.17 Noise temperature test curve

of Eqs. 3.36 and 3.37. Through the numerical analysis of the above test data, it can be seen that the scintillation noise is the main part of the structure noise at room temperature, but with the increase of temperature, mechanical thermal noise and electronic thermal noise gradually replace the scintillation noise and become the main part of the structure noise.

3.6 Chapter Summary

This chapter first introduces the equivalent electrical model of a single mass gyro (which is the same as that of another mass gyro) and the interface circuit of loop tube detection. Meanwhile, the anti-noise characteristics and working principle of loop tube interface circuit are introduced in detail based on the model. Again, according to the structure of gyroscope in the second chapter the ideal equation of motion, established the gyro in Simulink environment of the structure of the system model, and introduces the mechanical thermal noise and driving mode non-ideal factors, such as in Coriolis force to the quadrature error for subsequent chapters, testing modal force feedback, test modal frequency tuning provides the content such as the simulation platform. Finally, the silicon micromechanical gyroscope is designed to drive the modal self-excited drive closed loop, was analyzed by using the "average" the stability of the closed loop, and based on the gyro meter model of drive circuit are simulated and the temperature experiment, verified the drive circuit between -40 °C to 60 °C—driven displacement can be set at a very good control 1.56 μm to 1.57 μm, the drive loop has good amplitude stability. Finally, this paper introduces the noise in dual-mass linear vibration silicon-based MEMS gyroscope structure, and quantify the noise with the experiment.

Chapter 4
Dual-Mass Linear Vibration Silicon-Based MEMS Gyroscope Quadrature Error Correction Technology and Optimization

4.1 Chapter Introduction

For the ideal structure of gyro and monitoring circuit, Coriolis signal and circuit noise exist in the output, its character is only determined by the temperature drift, but it can be known from Figs. 4.1, 4.2, 4.3, 4.4, 4.5, 4.6, 4.7, 4.8, 4.9, 4.10, 4.11, 4.12, 4.13, 4.14, 4.15 gyro structure are greatly influenced by the processing error, these unfavorable factors to a large extent determine the zero drift and characteristic of the output signal. Reasonable analysis of the expression of machining error in engineering and quantification of each component of the output signal can be more targeted to design the relevant loop and improve the performance of gyro.

A large number of literatures have pointed out that quadrature error is the most influential factor of gyro performance caused by machining error. In this chapter, the causes and manifestations of quadrature error in dual-mass gyro are deeply analyzed and studied. Based on the correct understanding of its influence on the output signal, relevant correction methods are put forward for the main links, and the system model is analyzed and simulated in detail. Through relevant experiments on actual circuits, the optimal scheme for quadrature correction of dual-mass linear vibration silicon-based MEMS gyroscope is explored, which lays a solid foundation for the follow-up work.

4.2 Structural Error

In silicon micromechanical gyroscopes, structural errors are one of the most important error sources. Common structural errors include [123,124]: primary errors in design, such as modal overlap, motion coupling, mechanical parameters changing with temperature, etc., which can only be restricted by scientific and rigorous design schemes. Errors in the process of machining [20, 125–127], such as the deviation

H. Cao, *Dual-Mass Linear Vibration Silicon-Based MEMS Gyroscope*,
https://doi.org/10.1007/978-981-19-9247-6_4

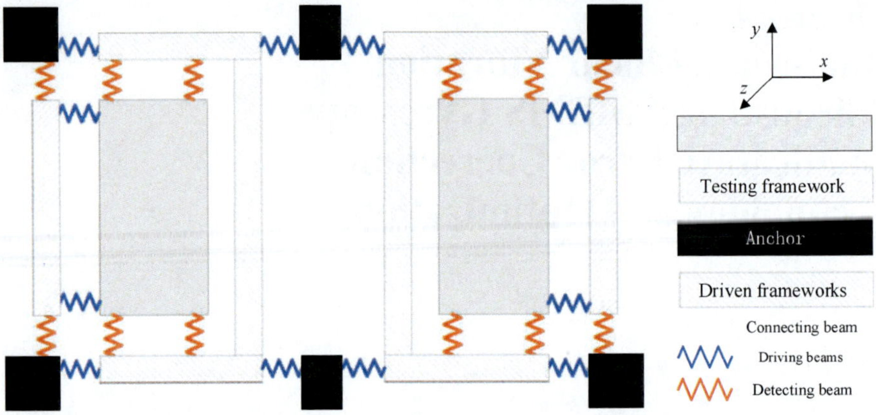

Fig. 4.1 Mass block-spring equivalent diagram of dual-mass linear vibration silicon-based MEMS gyroscope structure

between the size and position of the processed structure and the design value, asymmetry of the support beam, loss of comb teeth, deflection and unequal spacing [54, 55], will lead to unequal elasticity, damping asymmetry and mass imbalance of the structure [124, 128]. The structure error of silicon micromechanical gyroscope will affect the zero bias stability and repeatability of the gyroscope output.

Due to the limitation of silicon microstructure forming once and difficulty in subsequent nulling, the relative error generated by machining error is very large, which makes it one of the most important errors affecting the static performance of gyroscope in practical engineering.

4.2.1 Quadrature Error in Dual-Mass Linear Vibration Silicon-Based MEMS Gyroscope

The unequal elasticity generated by the machining error will make the Motion of The Coriolis mass with the same frequency but 90° phase difference with the Coriolis motion, which is usually called quadrature motion, and the error generated by the motion is called quadrature error.

4.2.1.1 The Generation Principle of Quadrature Error

The deviation of structure parameters caused by machining error will lead to the inconsistency between the drive and test elastic spindle and the design position (inertia spindle) after machining, or even the two axes are not completely vertical, which directly affects the force and motion of the drive and test mode in gyro structure. In order to clarify the influence of quadrature motion of dual-mass linear vibration

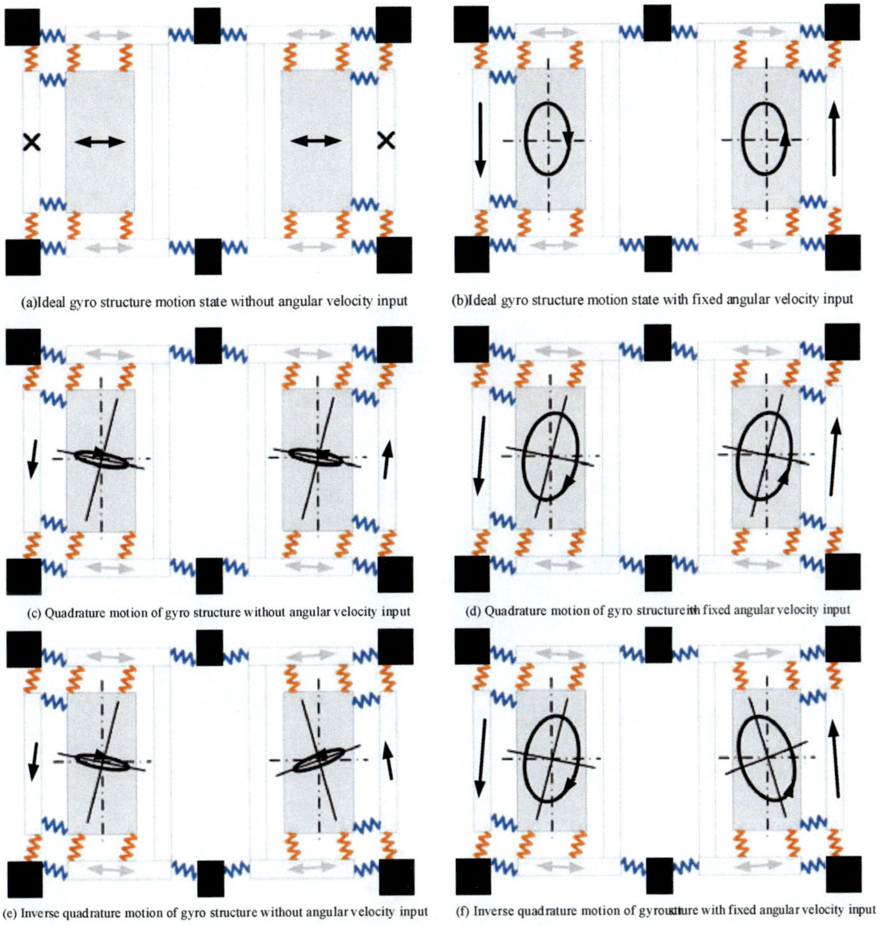

Fig. 4.2 Motion state diagram of ideal model of dual-mass linear vibration silicon-based MEMS gyroscope structure and model with quadrature error

Fig. 4.3 Calculation model of quadrature coupling stiffness

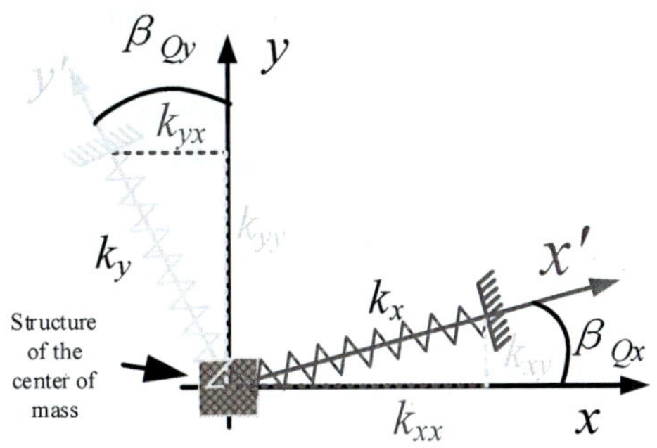

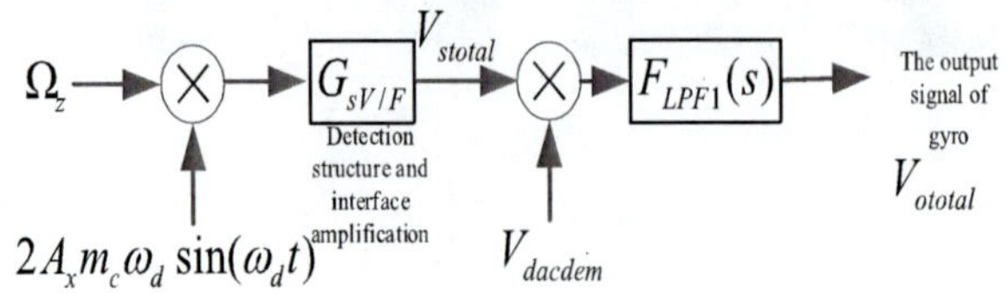

Fig. 4.4 Schematic diagram of gyro sensing open-loop

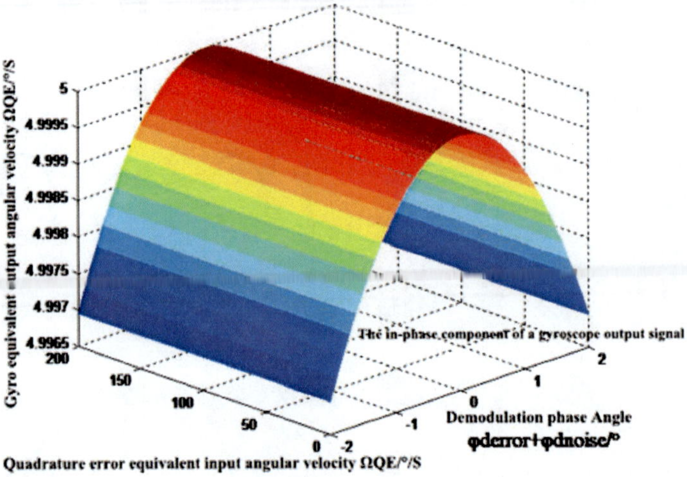

Fig. 4.5 Relationship between the same confidence signal and the gyroscope output signal

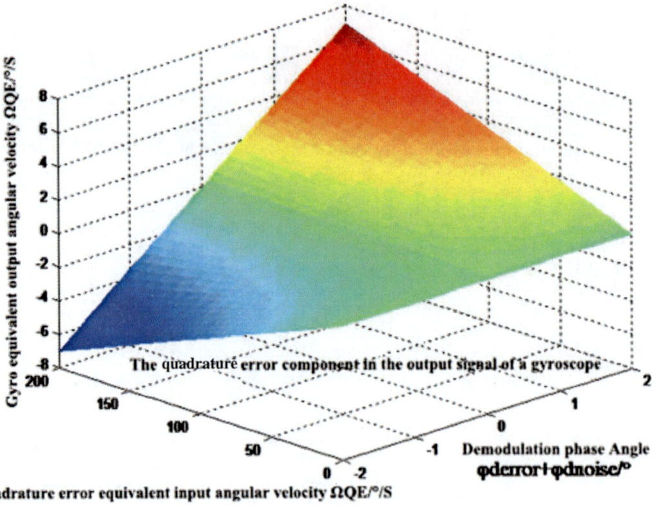

Fig. 4.6 Relationship between quadrature component and gyroscope output signal

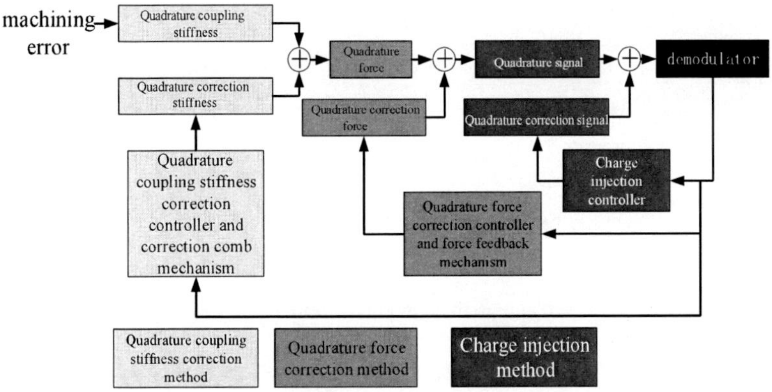

Fig. 4.7 Model of gyro structure system with coupling module

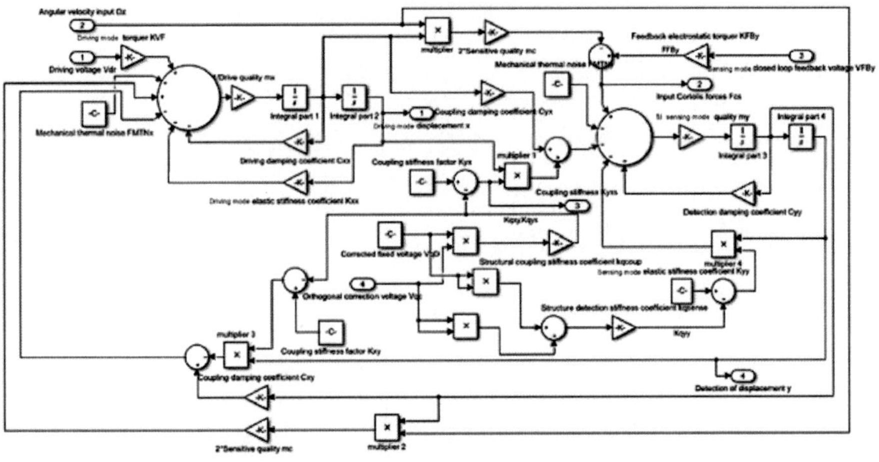

Fig. 4.8 Model of gyro structure system with coupled modules

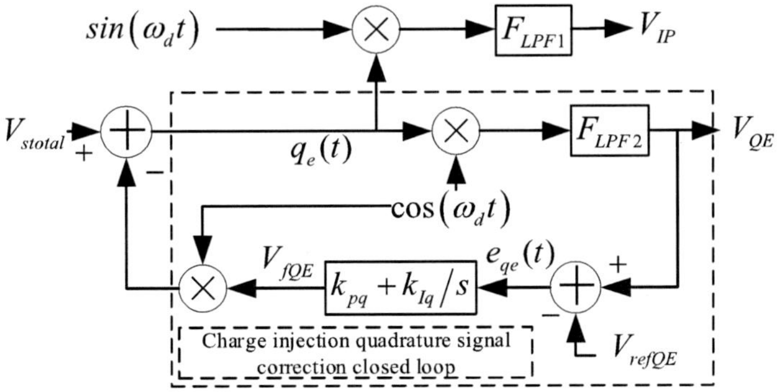

Fig. 4.9 Schematic diagram of charge injection quadrature nulling

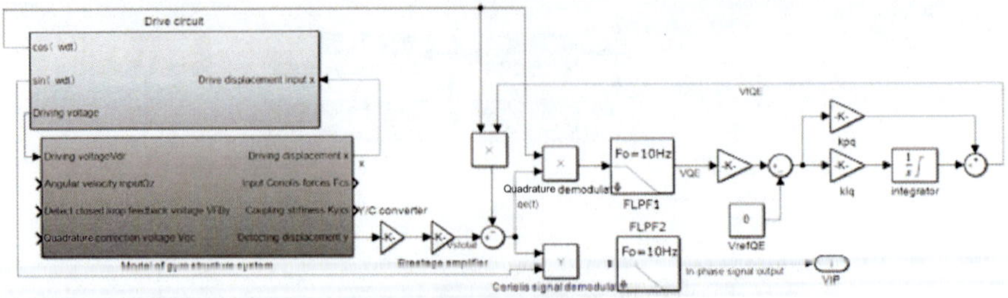

Fig. 4.10 Simulation diagram of charge injection quadrature nulling system

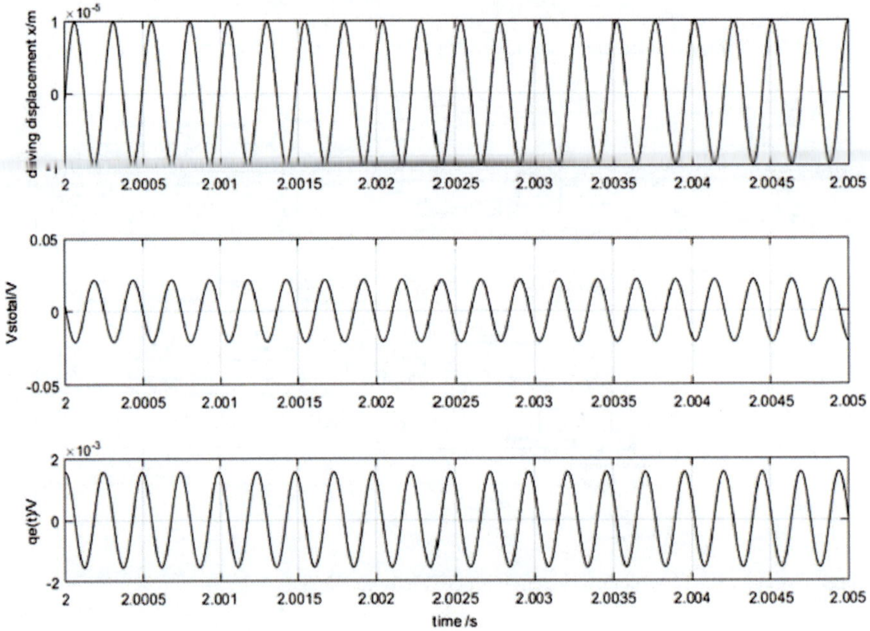

Fig. 4.11 The phase comparison results of charge injection method

silicon-based MEMS gyroscope structure, the structure in Fig. 2.5 is simplified in this section, and only the mass block-spring system is retained. Its equivalent figure is shown in Fig. 4.1. The motion of the modes in the gyro structure under ideal conditions without angular velocity input is shown in Fig. 4.2a, and its motion with fixed angular velocity input is shown in Fig. 4.2b. The modal motion of the non-angular velocity input and the fixed angular velocity input with quadrature motion are shown in Fig. 4.2c–f respectively. Where, (c) and (d) are the results of two mass blocks affected by processing errors in the same direction, and (e) and (f) are the results of two mass blocks affected by processing errors in the opposite direction.

As can be seen from the figure above, both the driving and sensing spring spindles in the ideal model coincide with the inertial spindles (X and Y axes, dotted in the

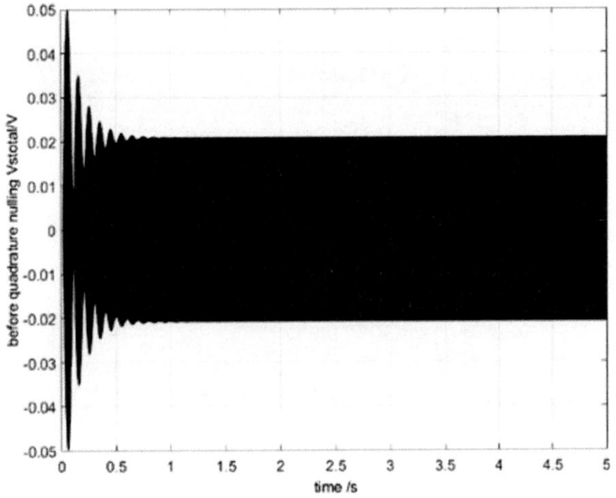

Fig. 4.12 V_{stotal} simulation results of charge injection method

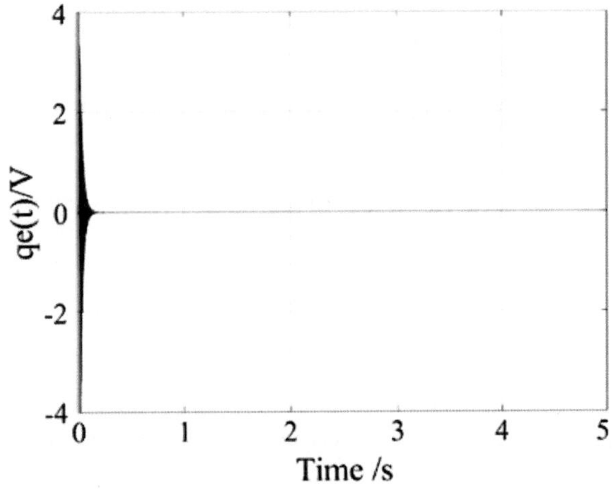

Fig. 4.13 $q_e(t)$ simulation results of charge injection method

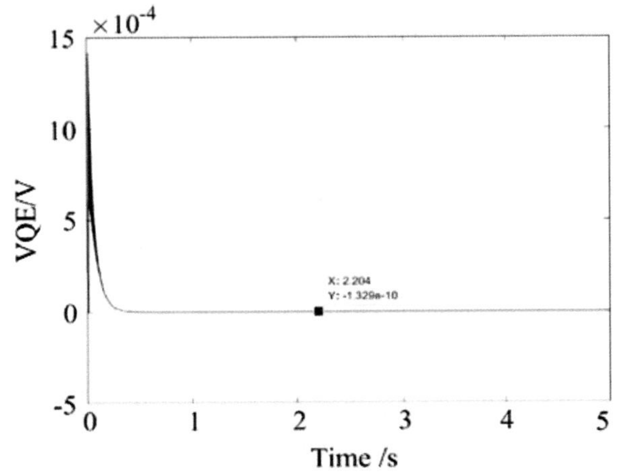

Fig. 4.14 V_{QE} simulation results of charge injection method

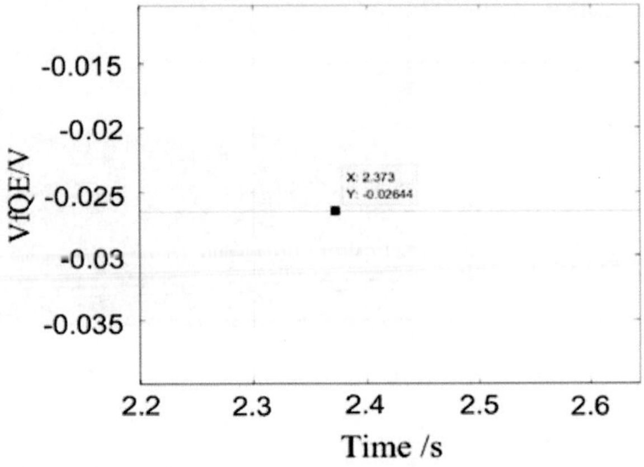

Fig. 4.15 V_{fQE} simulation results of charge injection method

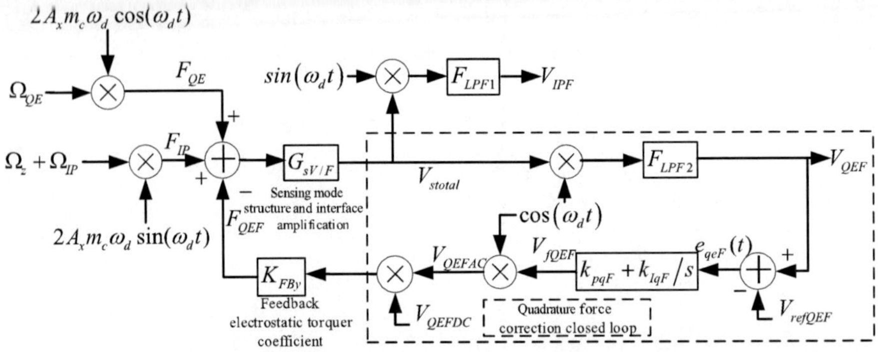

Fig. 4.16 Block diagram of quadrature force correction method

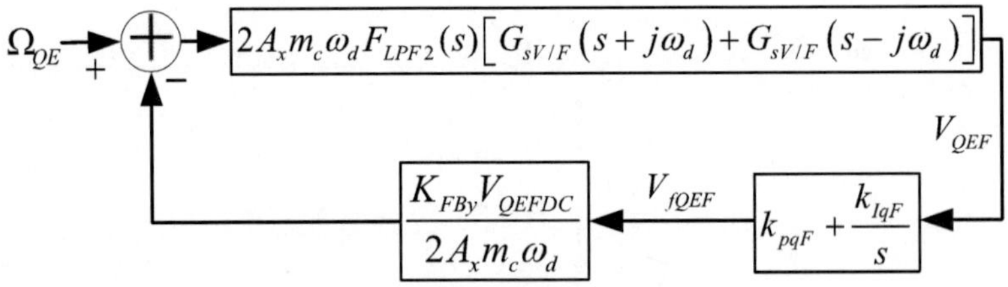

Fig. 4.17 Simplified system block diagram by quadrature force correction method

figure). In figure (a), when there is no angular velocity input, the driving mode checks that the frame is in a static state (represented by " × ") at work, and the driving frame and Coriolis mass only move in the X direction. When angular velocity inputs, the driving and sensing modes move along X and Y axes, respectively, as shown in Fig. (b), and the Coriolis mass moves along a positive ellipse that recombines

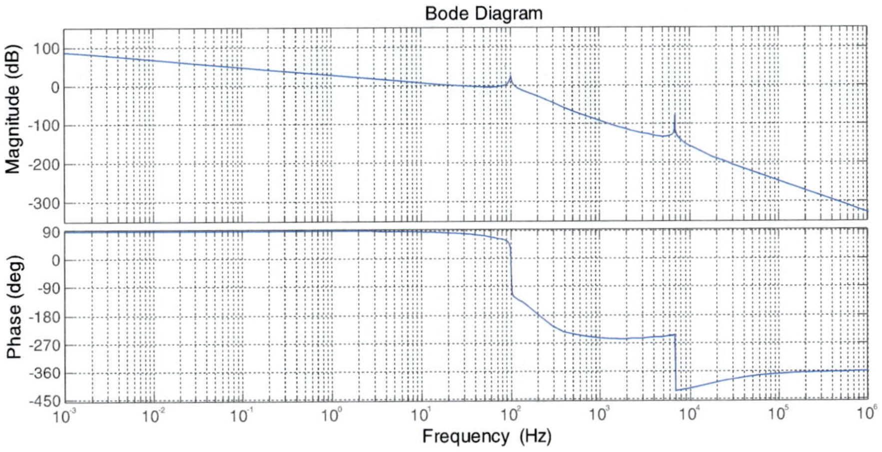

Fig. 4.18 Open-loop Bode diagram of quadrature force correction system

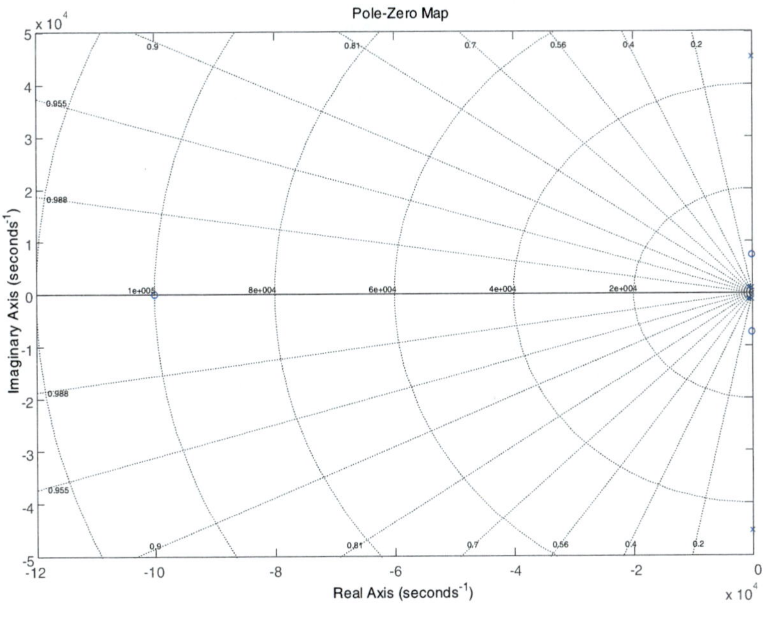

Fig. 4.19 Distribution of zero and pole in closed loop of quadrature force correction system

with X and Y axes, respectively, with long and short axis. These two are the ideal state, but in practical application, can make the structure of the gyro spring spindle machining error (figure of fine solid black line) and the inertial principal axis has certain direction angle (easy to express, a book with the quadrature error Angle "to represent the angle"), quadrature movement caused by the angle, under the influence of the perpendicular movement, the motion of Coriolis mass is an elliptical motion

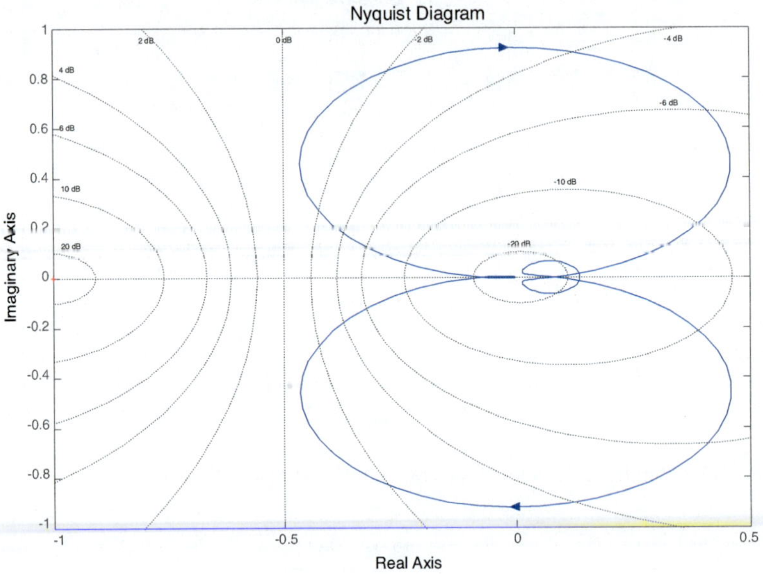

Fig. 4.20 Closed loop Nyquist curve of quadrature force correction system

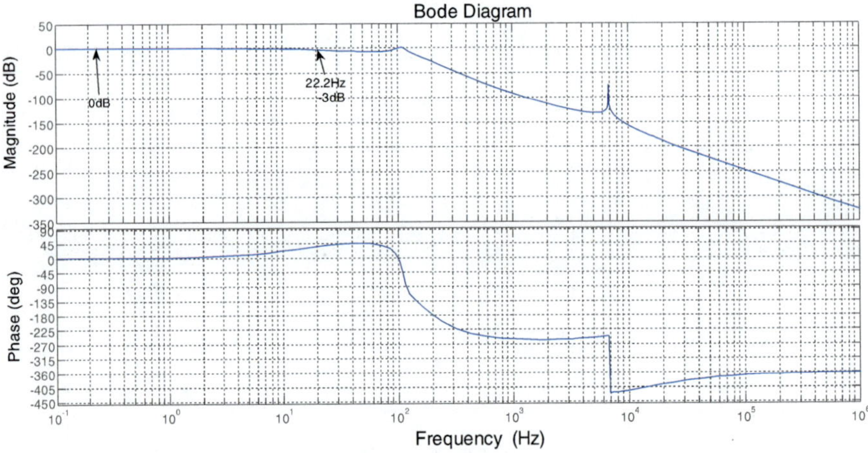

Fig. 4.21 Closed loop Bode diagram of quadrature force correction system

in which the long and short axes do not coincide with the X and Y axes. In addition, it is difficult to meet the complete symmetry of the two mass blocks in the process of machining, so the included angle of quadrature error generated by the left and right mass blocks is also difficult to be exactly the same, or even reverse. As shown in Fig. 4.2c and e, even without angular velocity input, the detection frame will have a certain displacement under the influence of quadrature motion, and the displacement amplitude is determined by the quadrature motion. When fixed input angular velocity,

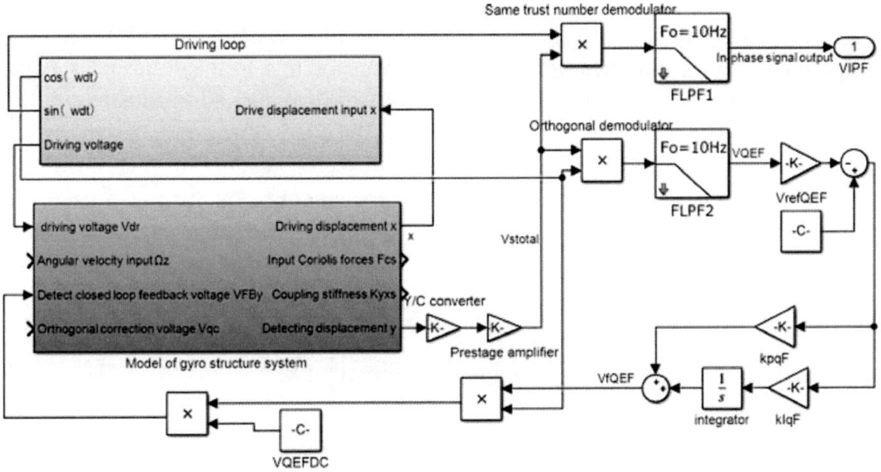

Fig. 4.22 System simulation diagram of quadrature force correction method

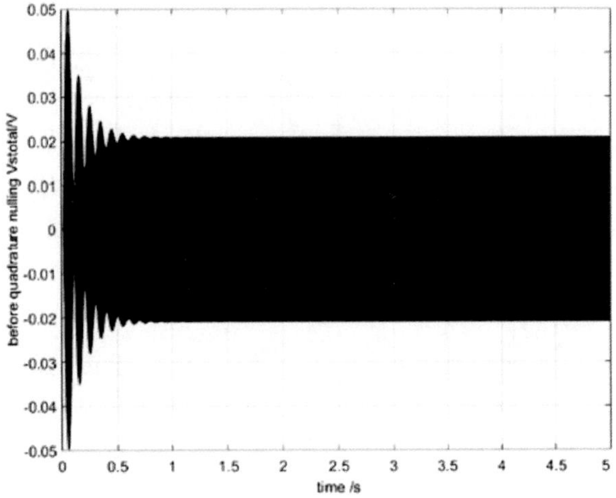

Fig. 4.23 V_{stotal} simulation results before quadrature force correction

such as (d) and (f), sensing mode amplitude also increases, at the same time, the direction of displacement detection coupled to driving mode, but due to the driving range is larger displacement of the coupling can be neglected when the small angular rate input, but when the input angular rate is bigger, the adverse effects to the driving mode coupling games.

4.2.1.2 Quadrature Error Coupling Stiffness

The mass block-spring system is further simplified, and the equivalent driving and sensing axes can be obtained by vector superposition of the spring in the processed structure along the design axis, as shown in Fig. 4.3. In the figure, X and Y axes are the inertia axes of the modes (design axes), corresponding to the solid black lines with arrows in Fig. 4.2. X' and Y' are respectively the driving and sensing axes after processing. β_{Qx} and β_{Qy} are the included angles of the quadrature errors of the driving and sensing axes respectively, which can be positive or negative in relation to the machining errors. k_x and k_y are the actual stiffness generated by the machined drive and test beam, acting on the X' and Y' axes respectively. The actual forces of k_x and k_y on X and Y axes can be obtained by projection method and "stiffness principal axis principle" method [70, 94, 129]:

$$\begin{bmatrix} F_{kx} \\ F_{ky} \end{bmatrix} = \begin{bmatrix} \cos\beta_{Qx} & -\sin\beta_{Qy} \\ \sin\beta_{Qx} & \cos\beta_{Qy} \end{bmatrix} \begin{bmatrix} F'_{kx} \\ F'_{ky} \end{bmatrix} = \begin{bmatrix} k_x\cos\beta_{Qx} & -k_y\sin\beta_{Qy} \\ k_x\sin\beta_{Qx} & k_y\cos\beta_{Qy} \end{bmatrix} \begin{bmatrix} x' \\ y' \end{bmatrix} \tag{4.1}$$

Coordinate transformation exists at the same time:

$$\begin{bmatrix} x' \\ y' \end{bmatrix} = \begin{bmatrix} \cos\beta_{Qx} & \sin\beta_{Qx} \\ -\sin\beta_{Qy} & \cos\beta_{Qy} \end{bmatrix} \begin{bmatrix} x \\ y \end{bmatrix} \tag{4.2}$$

Substitute the above equation into Eq. (4.1) to obtain:

$$\begin{aligned} \begin{bmatrix} F_{kx} \\ F_{ky} \end{bmatrix} &= \begin{bmatrix} k_x\cos\beta_{Qx} & -k_y\sin\beta_{Qy} \\ k_x\sin\beta_{Qx} & k_y\cos\beta_{Qy} \end{bmatrix} \begin{bmatrix} \cos\beta_{Qx} & \sin\beta_{Qx} \\ -\sin\beta_{Qy} & \cos\beta_{Qy} \end{bmatrix} \begin{bmatrix} x \\ y \end{bmatrix} \\ &= \begin{bmatrix} k_x\cos^2\beta_{Qx} + k_y\sin^2\beta_{Qy} & k_x\sin\beta_{Qx}\cos\beta_{Qx} - k_y\cos\beta_{Qy}\sin\beta_{Qy} \\ k_x\sin\beta_{Qx}\cos\beta_{Qx} - k_y\sin\beta_{Qy}\cos\beta_{Qy} & k_x\sin^2\beta_{Qx} + k_y\cos^2\beta_{Qy} \end{bmatrix} \begin{bmatrix} x \\ y \end{bmatrix} \end{aligned} \tag{4.3}$$

According to the content of the first matrix on the right of the above formula, suppose:

$$\begin{cases} k_{xx} = k_x\cos^2\beta_{Qx} + k_y\sin^2\beta_{Qy} \\ k_{xy} = k_x\sin\beta_{Qx}\cos\beta_{Qx} - k_y\cos\beta_{Qy}\sin\beta_{Qy} \\ k_{yx} = k_x\sin\beta_{Qx}\cos\beta_{Qx} - k_y\sin\beta_{Qy}\cos\beta_{Qy} \\ k_{yy} = k_x\sin^2\beta_{Qx} + k_y\cos^2\beta_{Qy} \end{cases} \tag{4.4}$$

In the above formula, k_{xx}, k_{yy}, k_{yx} and k_{xy} are the actual stiffness coefficients of the driving and sensing modes respectively, the stiffness coefficients of the driving mode coupling to the sensing mode and the stiffness coefficients of the sensing mode coupling to the driving mode. And $k_{yx} = k_{xy}$, the two-coupling stiffness is the expression form of quadrature error in gyro mechanics model, but the driving mode amplitude is much larger than the sensing mode amplitude, so the quadrature coupling force of the driving to sensing mode is large, while the coupling force of the sensing to driving mode can be ignored because it is much smaller than the driving force. In actual engineering, $\beta_{Qx} \approx \beta_{Qy}$, then:

$$k_{xy} = k_{yx} = \frac{k_x - k_y}{2} \sin 2\beta_{Qx} \tag{4.5}$$

Therefore, decreasing the difference of gyro mode design stiffness and decreasing the included angle of quadrature error can effectively reduce the quadrature coupling stiffness and restrain the influence of quadrature error.

4.2.2 Asymmetrical Damping

The unequal size and spacing of comb teeth caused by machining errors can lead to asymmetrical damping of gyro structure, which results in an angle between damping spindle of driving mode and inertial spindle of sensing mode, which can be set as γ_{Qx} and γ_{Qy} respectively. In addition, the uneven distribution of air and other media in gyro structure also leads to asymmetrical damping. Similar to the analysis of coupling stiffness, "damping spindle principle" is introduced to obtain:

$$\begin{aligned}\begin{bmatrix} F_{cx} \\ F_{cy} \end{bmatrix} &= \begin{bmatrix} c_x \cos\gamma_{Qx} & -c_y \sin\gamma_{Qy} \\ c_x \sin\gamma_{Qx} & c_y \cos\gamma_{Qy} \end{bmatrix} \begin{bmatrix} \cos\gamma_{Qx} & \sin\gamma_{Qx} \\ -\sin\gamma_{Qy} & \cos\gamma_{Qy} \end{bmatrix} \begin{bmatrix} \dot{x} \\ \dot{y} \end{bmatrix} \\ &= \begin{bmatrix} c_x \cos^2\gamma_{Qx} + c_y \sin^2\gamma_{Qy} & c_x \sin\gamma_{Qx}\cos\gamma_{Qx} - c_y \cos\gamma_{Qy}\sin\gamma_{Qy} \\ c_x \sin\gamma_{Qx}\cos\gamma_{Qx} - c_y \sin\gamma_{Qy}\cos\gamma_{Qy} & c_x \sin^2\gamma_{Qx} + c_y \cos^2\gamma_{Qy} \end{bmatrix} \begin{bmatrix} \dot{x} \\ \dot{y} \end{bmatrix}\end{aligned} \tag{4.6}$$

Also, suppose:

$$\begin{cases} c_{xx} = c_x \cos^2\gamma_{Qx} + c_y \sin^2\gamma_{Qy} \\ c_{xy} = c_x \sin\gamma_{Qx}\cos\gamma_{Qx} - c_y \cos\gamma_{Qy}\sin\gamma_{Qy} \\ c_{yx} = c_x \sin\gamma_{Qx}\cos\gamma_{Qx} - c_y \sin\gamma_{Qy}\cos\gamma_{Qy} \\ c_{yy} = c_x \sin^2\gamma_{Qx} + c_y \cos^2\gamma_{Qy} \end{cases} \tag{4.7}$$

In the above formula, c_{XX}, c_{yy}, c_{yx} and c_{xy} are the actual damping coefficients of the driving mode and the sensing mode respectively, the damping coefficients of the driving mode coupling to the sensing mode and the sensing mode coupling to the driving mode. $c_{yx} = c_{xy}$, and $\gamma_{Qx} \approx \gamma_{Qy}$, then:

$$c_{xy} = c_{yx} = \frac{c_x - c_y}{2} \sin 2\gamma_{Qx} \tag{4.8}$$

Because the frequency and phase of the force produced by gyroscopic coupling damping are the same as that of the Coriolis signal, the damping coupling factor cannot be eliminated in the circuit and will appear as part of the zero position of the gyroscope output. In practical engineering, because of the vaccum package, the damping coefficient of gyro is very small, and its influence on zero output can be ignored, but when the temperature changes, the coupling damping will change and lead to the change of zero of gyro, which can be optimized by temperature compensation technology. Combined with the data of the simulation model in the previous chapter, the coupled damping force is more than three to four orders of magnitude smaller than the coupled elastic force.

4.3 Components of the Gyroscope Output Signal

Theoretically, when there is a large difference in the frequency between the driving and sensing modes, the Coriolis signal and the driving voltage signal in the sensing mode channel are in the same frequency and phase, so most of the existing literature adopts the phase sensitive demodulation scheme. In this book, the output V_{dac} of the 90° phase shift link in the driving loop is taken as the demodulation benchmark. Therefore, under ideal circumstances, the output signal of gyroscope reflects the size of The Coriolis signal in the sensing mode channel, but in fact, due to the influence of the interference signal in the gyroscope structure and the phase shift in the circuit, the output signal often has a large error. Therefore, analyzing and quantifying the components of gyroscope output signal can suppress and compensate the possible errors and interference items, and then improve the performance of gyroscope. In order to facilitate the analysis of phase and amplitude characteristics of each component, this section adopts the open-loop detection loop. Figure 4.4 shows the simplified block diagram of the sensing loop, and the detailed analysis of the detection loop will be introduced in the next chapter. In the figure, $G_{sV/F}$ is the equivalent transfer function of amplitude change of gyro sensing mode structure and subsequent interface and amplification circuit. F_{LPF1} is a low-pass filter. Let $x(t) = A_x cos\ (\omega_{DT})$, $V_{dacdem} = V_{dacsin}\ (\omega_{DT} + \varphi_{derror} + \varphi_{dnoise})$, where φ_{derror} is demodulation phase angle error, φ_{dnoise} is demodulation phase angle noise, caused by electronic device drift and noise in the circuit. The signal V_{stotal} before the phase sensitive demodulation link of the gyroscope sensing loop circuit mainly contains components [62, 130, 131]: The Coriolis signal V_{sC}, the signal V_{sM} generated by mechanical coupling, the signal V_{sE} generated by electrical coupling, and the signal V_{sF} generated by force coupling. The four components are analyzed below. It should be pointed out that the phase shift of the force will occur when it passes through $G_{sV/F}$. In order to facilitate the analysis, the phase shift of this link is equivalent to φ_{derror} and φ_{dnoise} in the understanding adjustment reference.

4.3.1 Coriolis Force Signal

Coriolis signal V_{sC} is a signal detected by the gyroscope that is related to angular velocity. In V_{stotal}, it is represented as a signal modulated by the driving force AC signal:

$$V_{sC} = 2\Omega_z A_x \omega_d m_c G_{sV/F} \sin(\omega_d t) \tag{4.9}$$

As can be seen from the above equation, if the gyroscope detection structure and subsequent gain remain unchanged, the proportion of the Coriolis signal in the detection path can be improved by increasing the driving amplitude, increasing the driving mode resonant frequency and increasing the Coriolis mass, which is conducive to the detection of the Coriolis signal and the improvement of the SNR.

4.3.2 Mechanical Coupling Signal

Mechanical coupling signal V_{sM} is mainly composed of two parts, one is quadrature error V_{sMk} generated by stiffness coupling, and the other is V_{sMc} generated by damping coupling. According to the analysis in the previous section, it can be known that V_{sMk} and V_{sMc} are quadrature to and in phase with Coriolis signal respectively, which can be expressed as:

$$V_{sMk} = \alpha_{sMk} A_x G_{sV/F} \cos(\omega_d t) \tag{4.10}$$

$$V_{sMc} = \alpha_{sMc} A_x \omega_d G_{sV/F} \sin(\omega_d t) \tag{4.11}$$

In the formula, α_{smk} and α_{smc} are coefficients respectively, which are related to k_{yx} and c_{yx}. There are different mechanical coupling signals between different structures. The gyroscope with small mechanical coupling signal can be obtained by screening and machining the different structure protypes. In addition, due to V_{sMk} >> V_{sMc}, and it is pointed out in the previous section that the influence of V_{sMc} can be weakened through the method of structural vacuum, so this section mainly eliminates the influence of quadrature error V_{sMk}. The compensation signal can be added in the sensing loop circuit and the quadrature signal in the same amplitude phase reversal; The modes are detected by electrostatic force driving in reverse phase of quadrature signal. The coupling stiffness is compensated by negative static stiffness generated by additional compensating electrode. Quadrature signals are compensated by designing reasonable support beams.

4.3.3 Electrically Coupled Signals

Because both the driving electrode and the sensing electrode have one end connected to the Coriolis mass through the beam and comb frame, that is, they have the same common end, so the driving signal can be directly output to the sensing loop interface through the driving capacitor and sensing capacitor. Then electrically coupled signal V_{sE} can be expressed as:

$$V_{sE} = \alpha_{sE} V_{ac} \sin(\omega_d t + \theta_{sE}) \tag{4.12}$$

where α_{sE} is a constant, it is related to the amplitude of electrical coupling. θ_{sE} is the phase difference generated during electrical coupling. Differential detection can be used; Symmetrical, small stray capacitance design; The driving and sensing frequency separation methods reduce the influence of electrocoupling signal.

4.3.4 Force Coupled Signal

Due to machining error, the driving mode spindle and the sensing mode spindle cannot be completely quadrature, so the driving force will act on the sensing mode as well as the excitation mode. In general, the driving force is multiplied by a coefficient α_{sF} as the force coupling signal V_{sF} applied to the sensing mode:

$$V_{sF} = \alpha_{sF} F_d G_{sV/F} \sin(\omega_d t + \theta_{sF}) \tag{4.13}$$

Between it:

$$\theta_{sF} = -tg^{-1}\left(\frac{\omega_y \omega_d}{Q_y\left(\omega_y^2 - \omega_d^2\right)}\right) \tag{4.14}$$

When the design parameters of gyro structure given in Chap. 3 are substituted into the above equation, $\theta_{sF} \approx -0.5°$ can be obtained.

4.3.5 Silicon Based MEMS Gyroscope Output Signal

Through the analysis in the above sections, the signal in the sensing loop channel can be expressed as:

$$\begin{aligned} V_{stotal} &= V_{sC} + V_{sMk} + V_{sMc} + V_{sE} + V_{sF} \\ &= (2\Omega_z m_c + \alpha_{sMc}) A_x \omega_d G_{sV/F} \sin(\omega_d t) + \alpha_{sMk} A_x G_{sV/F} \cos(\omega_d t) \\ &+ \alpha_{sE} V_{ac} \sin(\omega_d t + \theta_{sE}) + \alpha_{sF} F_d G_{sV/F} \sin(\omega_d t + \theta_{sF}) \end{aligned} \tag{4.15}$$

Under the sequential action of multiplication demodulator and low-pass filter, the output signal V_{ototal} of the gyroscope is:

$$\begin{aligned} V_{ototal} &= (V_{stotal} V_{dacdem})\big|_{FLPF} \\ &= \frac{1}{2}[\alpha_{sF} F_d G_{sV/F} V_{dac} \cos(\varphi_{derror} + \varphi_{dnoise} - \theta_{sF}) + \alpha_{sMk} A_x V_{dac} G_{sV/F} \sin(\varphi_{derror} + \varphi_{dnoise}) \end{aligned}$$

$$+\alpha_{sE} V_{ac} V_{dac} \cos(\varphi_{derror} + \varphi_{dnoise} - \theta_{sE}) + (2\Omega_z m_c + \alpha_{sMc}) A_x \omega_d G_{sV/F} V_{dac} \cos(\varphi_{derror} + \varphi_{dnoise})] \quad (4.16)$$

Since $\varphi_{derror} + \varphi_{dnoise}$, θ_{sE} and θ_{sF} are all small angles, the above equation can be divided into sine and cosine components. The cosine component is called the in-phase component of The Coriolis signal because it contains the Coriolis signal. Sinusoidal components are mainly composed of quadrature signals. Further simplify the above equation:

$$\begin{aligned} V_{ototal} \approx & \frac{V_{dac}}{2} \{ \alpha_{sMk} A_x G_{sV/F} (\varphi_{derror} + \varphi_{dnoise}) \\ & + [\alpha_{sE} V_{ac} + \alpha_{sF} F_d G_{sV/F} + (2\Omega_z m_c + \alpha_{sMc}) A_x \omega_d G_{sV/F}] \} \end{aligned} \quad (4.17)$$

It can be seen from the above equation that the phase-sensitive demodulation method cannot suppress the influence of in-phase components, and the phase angle error and noise of demodulation will also introduce part of quadrature signals into the output signal.

4.3.6 Influence of Quadrature Error on Gyroscope Output Signals

In order to analyze and quantify the influence of quadrature signal on the output signal of gyroscope, $\Omega_z = 0$ in the formula (4.17) is used to quantify the amplitude of in-phase and quadrature components with the equivalent input angular velocity. To:

$$\Omega_{IP} = \frac{\alpha_{sE} V_{ac} + \alpha_{sF} F_d G_{sV/F} + \alpha_{sMc} A_x \omega_d G_{sV/F}}{2 m_c A_x \omega_d G_{sV/F}} \quad (4.18)$$

$$\Omega_{QE} = \frac{\alpha_{sMk}}{2 m_c \omega_d} \quad (4.19)$$

where, Ω_{IP} and Ω_{QE} are the equivalent input angular velocities of the same confidence signal and quadrature signal respectively, which are usually constant. In practical engineering, the typical value is several degrees /s, while the amplitude of the quadrature error is usually ten or even dozens of times of the same belief signal [62, 124], that is, if Ω_{QE} is tens to hundreds of degrees /s, as long as the demodulation phase angle is slightly deviated, the quadrature signal will bring great influence, so it is necessary to carry out thorough correction and compensation for the quadrature error. Figures 4.5 and 4.6 respectively describe the quantization influence of the in-phase and quadrature signal components in the output signal of the gyroscope, where the identical confidence sign is the constant value $\Omega_{IP} = 5°/s$, and the quadrature signal is the variable $\Omega_{QE} < 200°/s$, the demodulation phase angle error and noise

$\varphi_{derror} + \varphi_{dnoise}$ are between $-2°$ and $2°$. It can be seen that the size of quadrature signal does not affect the possession of the same confidence signal component in the gyroscope output signal. Within the range of demodulation phase angle change, Ω_O corresponding to the belief signal changes only 0.003°/s, while the quadrature error component changes more sharply with the demodulation phase angle change, Ω_O changes nearly 15°/s. Therefore, the suppression of quadrature error can effectively improve the performance of gyro.

4.4 Quadrature Error Correction

In a fully decoupled structure, quadrature errors are theoretically generated due to machining errors, so the correction of quadrature errors in the structure described in this book is usually carried out by the following methods [94]:

(1) Structural design and processing stage: simple structure can reduce the processing error, symmetrical structure, and with a large amount of residual thermal stress release structure (such as "U-beam"). Using high precision machining equipment and proper machining technology is beneficial to reduce machining error.
(2) Structural modification: fine, precise modification (e.g., laser modification) of machined silicon structures through a series of technical methods to correct machining errors within a relatively small range. At present, the application cost of this method is very high, and the structure can only be modified in a small range, which is not conducive to the production of large quantities of gyro structure. Moreover, individual modification schemes for different gyro structures are not the same, which also increases the application difficulty of this method.
(3) Compensation of quadrature signals in detection channels (charge injection method) [132]: This scheme mainly suppresses quadrature signals in sensing loop channels.
(4) Quadrature force correction method: because the quadrature error is generated by the coupled elastic force (quadrature force) acting on the sensing mode frame, the correction object of this method is the quadrature force of the sensing mode frame.
(5) Quadrature coupling stiffness correction method: this method uses quadrature correction comb teeth to generate electrostatic negative stiffness, so as to offset the coupling stiffness between driving and sensing modes, so as to fundamentally eliminate the quadrature error.

Among the above methods, except for the first method, all the other methods are applied after the structure is processed and formed. However, the second method has obvious limitations. In this section, the other three methods are analyzed in detail, and the three methods are studied and compared in depth with the corresponding experimental results. Figure 4.7 shows the relationship between the three methods.

4.4.1 System Model with Coupling Stiffness and Damping

Combining the previous work in this book with the gyro structural system model in Chap. 3, coupled stiffness and damping modules that generate quadrature errors and in-phase components are added. Meanwhile, in order to facilitate the analysis below, force-feedback comb module and quadrature correction comb module are also configured in the model, so the dynamic equation of gyro structure is:

$$
\begin{bmatrix} m_x\ddot{x} \\ m_y\ddot{y} \end{bmatrix} + \begin{bmatrix} c_{xx} & c_{xy} - 2m_c\Omega_z \\ c_{yx} + 2m_c\Omega_z & c_{yy} \end{bmatrix}\begin{bmatrix} \dot{x} \\ \dot{y} \end{bmatrix} + \begin{bmatrix} k_{xx} + k_{qxx} & k_{xy} + k_{qxy} \\ k_{yx} + k_{qyx} & k_{yy} + k_{qyy} \end{bmatrix}\begin{bmatrix} x \\ y \end{bmatrix} = \begin{bmatrix} F_{dx} \\ F_{FBy} \end{bmatrix} - \begin{bmatrix} F_{MTNx} \\ F_{MTNy} \end{bmatrix} \tag{4.20}
$$

In the above formula, k_{q**} is the electrostatic negative stiffness generated by the output of quadrature correction; F_{FBy} is the feedback force of the feedback comb tooth. According to the above formula, an improved gyro structure system model can be obtained, as shown in Fig. 4.8 [120, 133, 134]. The parameters of coupling stiffness, damping and feedback torquer are shown in Table 4.1. In addition, related parameters of k_{q**} will be introduced in subsequent chapters.

4.4.2 Charge Injection Method

According to the method described in Fig. 4.7, charge injection method is to add voltage signals with the same frequency and opposite phase of the quadrature signal in the sensing loop channel to cancel the quadrature signal. Since charge injection targets electrical signals in circuits, there is no need to add additional calibration or force feedback electrodes to the gyro structure. The advantage of this method is to reduce the need for comb teeth and simplify the complexity of structure design, which is beneficial to reduce the difficulty of machining and improve the yield of finished products.

Table 4.1 Simulation parameters of gyroscopic structure system model coupling module

Parameter names	Parameter value
Coupling stiffness factor k_{xy}, k_{yx}	0.1719 N/m
Coupling damping coefficient c_{xy}, c_{yx}	1.29×10^{-7} N/m/s
The input angle rate corresponding to the coupling stiffness Ω_{QE}	200°/s

4.4.2.1 Working Principle of Charge Injection Method

Because the quadrature signals of different gyro structure protypes are different, it is required that the method can form a closed-loop system in the application process and independently complete the detection and suppression of quadrature signals, so as to enhance the universality of the circuit [132, 135–137]. Figure 4.9 shows the conceptual diagram.

Where, the dotted line box is the closed-loop compensation loop of quadrature signal nulling, and V_{refQE} is the reference in the quadrature nulling loop, which is usually set to zero. If the influence of θ_{sE} and θ_{sF} is ignored in Eq. (4.15), then:

$$\begin{aligned} V_{stotal} = & \left[(2\Omega_z m_c + \alpha_{sMc})A_x\omega_d G_{sV/F} + \alpha_{sE}V_{ac} + \alpha_{sF}F_d G_{sV/F}\right]\sin(\omega_d t) \\ & + \alpha_{sMk}A_x G_{sV/F}\cos(\omega_d t) \end{aligned} \tag{4.21}$$

Besides it, there are:

$$V_{QE}(t) = [q_e(t)\cos(\omega_d t)]\big|_{F_{LPF2}} = \left[\left(V_{stotal}(t) - V_{fQE}(t)\cos(\omega_d t)\right)\cos(\omega_d t)\right]\big|_{F_{LPF2}} \tag{4.22}$$

$$e_{qe}(t) = V_{QE}(t) - V_{ref\,QE}(t) \tag{4.23}$$

$$V_{fQE}(t) = e_{qe}(t)k_{pq} + k_{Iq}\int_0^t e_{qe}(t)dt \tag{4.24}$$

After substituting Eq. (4.21) into Eq. (4.22), ignoring demodulation phase angle error and noise and considering the role of low-pass filter:

$$V_{QE}(s) = \frac{\alpha_{sMk}A_x G_{sV/F}(s) - V_{fQE}(s)}{2} = \frac{A_{QE}(s) - V_{fQE}(s)}{2} \tag{4.25}$$

After combining the Laplace transform of Eqs. (4.23) and (4.24):

$$\frac{V_{QE}(s)}{A_{QE}(s)} = \frac{1}{2 + k_p + {}^{k_I}\!/_{s}} \tag{4.26}$$

The above equation has only one pole $-k_I/(2 + K_P)$, so it can be seen that the quadrature nulling closed-loop system with the output signal of gyro structure as the input signal and V_{QE} as the output signal is stable. When the system is static, $s = 0$, we have $k_P + k_I/s >> 2$, then:

$$A_{QE}(s) \approx V_{QE}(s)(k_p + {}^{k_I}\!/_{s}) = V_{fQE}(s) \tag{4.27}$$

So, the cosine component of q_e (t) is:

$$q_{e\cos}(t) = \left[A_{QE}(t) - V_{fQE}(t)\right]\cos(\omega_d t) \approx 0 \tag{4.28}$$

It can be seen from the above equation that charge injection method can effectively cancel the quadrature signals in the detection path under the stable state of the system.

4.4.2.2 Charge Injection System Simulation

According to the working principle of charge injection method introduced in the previous section, a simulation model is established in the Simulink environment, as shown in Fig. 4.10. The gyro structure system model is Shown in Fig. 4.8, and the driving loop contains the content in Fig. 3.9. Meanwhile, V_{stotal}, $q_e(t)$, V_{QE} and V_{fQE} are monitored and compared with the phase of driving displacement x, as shown in Figs. 4.11, 4.12, 4.13, 4.14, 4.15.

The figure above is the simulation curve between 2 s and 2.005 s of system operation. It can be seen from the figure that the main component of V_{stotal} is quadrature signal, and its peak value is about 20 mV, with a 180° difference between phase and driving displacement. The main component of Q_E signal after calibration is identical signal (90° phase difference with quadrature signal), and its peak value is less than 2 mV. The closed loop system based on charge injection method basically eliminates the quadrature signals in the detection loop.

It can be seen from Figs. 4.12, 4.13, 4.14, 4.15 that each link in the system can reach steady state in a very short time, and the amplitude of quadrature signal V_{QE} in the detection channel is stable at the level of nV, which is basically eliminated. The correction voltage V_{fQE} of the quadrature signal shows a fixed value, which is related to the quadrature signal in V_{stotal}. The results are consistent with the theoretical analysis in the previous section.

4.4.3 *Quadrature Force Correction Method*

In this method, the influence of quadrature force on the sensing mode frame is offset by applying the electrostatic force with the same frequency and opposite phase of quadrature force on the sensing mode feedback comb teeth [138–140]. The method requires the coordination between the external control circuit and the sensing mode feedback comb in the structure, and the control parameters and structural parameters need to be coordinated to achieve the optimal control state.

4.4.3.1 Working Principle of Quadrature Force Nulling Method

The quadrature error correction force is generated by sensing mode feedback sliding mode comb (its specific structure and parameter analysis in Chap. 5), the second chapter introduces the comb teeth produce electrostatic force need ac and dc voltage superposition, this section utilized the DC fixed and AC changeable control method and reference section on the mode of AGC loop. Figure 4.16 shows the working principal block diagram of quadrature force correction method. In the figure, the closed loop of quadrature force correction is in the dotted box. Where, F_{QE}, F_{IP} and F_{QEF} are quadrature force, Coriolis equal force and quadrature feedback force respectively. V_{QEFAC} and V_{QEFDC} are the AC and DC components of the calibration voltage respectively. V_{refQEF} is the reference voltage of the control point. As the electrostatic force is determined by the product of V_{QEFAC} and V_{QEFDC}, it is represented as the product in Fig. 4.16. According to the system block diagram shown in the figure above, the equation can be obtained:

$$F_{QE} = 2\Omega_{QE} A_x m_c \omega_d \cos(\omega_d t) \tag{4.29}$$

$$F_{IP} = 2(\Omega_z + \Omega_{IP}) A_x m_c \omega_d \sin(\omega_d t) \tag{4.30}$$

$$F_{QEF} = K_{FBy} V_{fQEF} V_{QEFDC} \cos(\omega_d t) \tag{4.31}$$

$$\begin{aligned} V_{stotal} =& \left(F_{QE} + F_{IP} - F_{QEF}\right) G_{sV/F} \\ =& \left[\left(2\Omega_{QE} A_x m_c \omega_d - K_{FBy} V_{fQEF} V_{QEFDC}\right) \cos(\omega_d t) \right. \\ & \left. + 2(\Omega_z + \Omega_{IP}) A_x m_c \omega_d \sin(\omega_d t)\right] G_{sV/F} \end{aligned} \tag{4.32}$$

4.4.3.2 Analysis of Static and Dynamic Characteristics of the System by Quadrature Force Nulling Method

This section mainly analyzes the quadrature and quadrature feedback signals in the loop, and takes the quadrature force equivalent input angle rate Ω_{QE} as the system input variable, and the feedback force control voltage V_{fQEF} as the system output. Take the Laplace transform of Eqs. (4.29) and (4.31):

$$F_{QE}(s) = A_x m_c \omega_d \left[\Omega_{QE}(s + j\omega_d) + \Omega_{QE}(s - j\omega_d)\right] \tag{4.33}$$

$$F_{QEF}(s) = \frac{1}{2} K_{FBy} V_{QEFDC} \left[V_{fQEF}(s + j\omega_d) + V_{fQEF}(s - j\omega_d)\right] \tag{4.34}$$

After gyro sensing mode transfer function, demodulator and low-pass filter (ignoring high-frequency components), there are:

$$\begin{aligned} V_{QEF}(s) =& \left\{\left[F_{QE}(s + j\omega_d) - F_{QEF}(s + j\omega_d)\right] G_{sV/F}(s + j\omega_d) \right. \\ & \left. + \left[F_{QE}(s - j\omega_d) - F_{QEF}(s - j\omega_d)\right] G_{sV/F}(s - j\omega_d)\right\} \big|_{F_{LPF2}} F_{LPF2}(s) \\ =& \left(2\Omega_{QE}(s) A_x m_c \omega_d - K_{FBy} V_{fQEF}(s) V_{QEFDC}\right) \end{aligned}$$

$$\times F_{LPF2}(s)\big[G_{sV/F}(s + j\omega_d) + G_{sV/F}(s - j\omega_d)\big] \tag{4.35}$$

where, $F_{LPF2}(s)$ is the transfer function of the low-pass filter, and $G_{sV/F}$ correlation function is:

$$\begin{aligned} &G_{sV/F}(s + j\omega_d) + G_{sV/F}(s - j\omega_d) \\ &= \frac{K_{YC}K_{pre}}{m_y}\left(\frac{1}{(s + j\omega_d)^2 + \frac{\omega_y}{Q_y}(s + j\omega_d) + \omega_y^2} + \frac{1}{(s - j\omega_d)^2 + \frac{\omega_y}{Q_y}(s - j\omega_d) + \omega_y^2}\right) \\ &= \frac{\frac{2K_{YC}K_{pre}}{m_y}\left(s^2 + \frac{\omega_y}{Q_y}s + \omega_y^2 - \omega_d^2\right)}{s^4 + \frac{2\omega_y}{Q_y}s^3 + \left(\frac{\omega_y^2}{Q_y^2} + 2\omega_d^2 + 2\omega_y^2\right)s^2 + 2\left(\frac{\omega_y^3}{Q_y} + \frac{\omega_d^2\omega_y}{Q_y}\right)s + \omega_d^4 + \omega_y^4 - 2\omega_d^2\omega_y^2 + \frac{\omega_y^2\omega_d^2}{Q_y^2}} \end{aligned} \tag{4.36}$$

If $V_{refQEF} = 0$, then

$$V_{fQEF}(s) = V_{QEF}(s)\left(k_{pqF} + \frac{k_{IqF}}{s}\right) \tag{4.37}$$

Substituting the above equation into Eq. (4.35), we can get:

$$\frac{V_{fQEF}(s)}{\Omega_{QE}(s)} = \frac{2A_x m_c \omega_d F_{LPF2}(s)\big[G_{sV/F}(s + j\omega_d) + G_{sV/F}(s - j\omega_d)\big]\left(k_{pqF} + \frac{k_{IqF}}{s}\right)}{1 + K_{FBy}V_{QEFDC}F_{LPF2}(s)\big[G_{sV/F}(s + j\omega_d) + G_{sV/F}(s - j\omega_d)\big]\left(k_{pqF} + \frac{k_{IqF}}{s}\right)} \tag{4.38}$$

Equation (4.38) is transformed into a unit negative feedback system, and the simplified block diagram of the quadrature force feedback correction method is shown in Fig. 4.17. According to Eq. (4.38), when the quadrature error equivalent input angle rate Ω_{QE} is fixed, $s = 0$, then the feedback control voltage is:

$$V_{fQEF} = \frac{2A_x m_c \omega_d}{K_{FBy}V_{QEFDC}}\Omega_{QE} \tag{4.39}$$

The above equation shows that the control voltage is directly proportional to Ω_{QE}, and V_{fQEF} has a good linear relationship with Ω_{QE} when all factors are stable. But when factors such as temperature change, the driving frequency changes accordingly, so V_{fQEF} will shift even if Ω_{QE} remains constant. To further verify the dynamic characteristics of the control system, dynamic analysis is performed on Fig. 4.17. Table 4.2 lists the parameters of each link in the system. By replacing the data in the table into Fig. 4.17, the Bode diagram of the system with open loop can be obtained, as shown in Fig. 4.18. It can be seen from the figure that the amplitude margin and phase angle margin of the system are about 21 dB and 89° respectively. After the system is closed, the zero-pole distribution can be obtained, as shown in Fig. 4.19.

Figure 4.20 shows that all poles (as shown by "×") are on the negative half cycle of the real axis; In addition, the closed loop Nyquist curve of the system is shown in

Table 4.2 Model parameters of each link of quadrature force correction system

Parameter name	Parameter value
Sensing mode displacement—capacitance conversion coefficient K_{YC}	1.501×10^{-9}F/m
Sensing mode interface capacitance–voltage conversion coefficient K_{pre}	-8.374×10^{11} V/F
Feedback electrostatic torquer K_{FBy}	1.487×10^{-7} N/V
Calibrate dc component of voltage V_{QEFDC}	5 V
Low pass filter F_{LPF2} cut-off frequency	200 Hz
Low pass filterF_{LPF2} quality factor	1
Low pass filter F_{LPF2} gain	−10
PI Controller proportionality k_{pqF}	0.01
PI Integral coefficient of controller k_{IqF}	1000

Fig. 4.20, and the reference point (−1, 0) is not surrounded by the curve. The above two criteria prove that the system is stable.

Figure 4.21 is the closed-loop Bode diagram of the quadrature force orrection system, in which the bandwidth of the system is about 12.5 Hz. Due to the small and slow variation of the quadrature force, the dynamic characteristics of the above system can fully meet the requirements of practical engineering. In addition, the amplitude feature in the figure of two rush for driving and sensing modes frequency phase impairment (low frequency peak position) and the addition of frequency value (high frequency peak location), it shows that when the frequency of the quadrature force equivalent input angular velocity Ω_{QE} for quadrature signal when two modal frequency offset is the largest, and usually Ω_{QE} as constant value or slow change. Therefore, in the frequency tuning state (driving and detecting natural frequencies are equal), the quadrature signal is the maximum value, which can be used as a judgment method of frequency tuning control.

4.4.3.3 Quadrature Force Correction System Time Domain Simulation

Combined with the gyro structural system model and the correction scheme proposed in the previous section, a closed-loop system is built in the Simulink environment, as shown in Fig. 4.22. Observe V_{QEF} and V_{fQEF} after nulling to judge the stability of the system. In addition, the state of V_{stotal} signal before and after correction is monitored, and the phase comparison is made with the driving displacement signal x.

It can be seen from Figs. 4.23, 4.24 and 4.25 that under the action of quadrature force correction method, the quadrature signal of V_{stotal} is basically eliminated, which proves that the quadrature force received by the sensing mode frame is basically canceled. At the same time, the sensing frame is still affected by the Coriolis signal in-phase force such as coupling damping force. It is worth noting that the amplitude peak of the same confidence sign in Fig. 4.25 is about 0.5 mV, which is smaller than that in Fig. 4.14 of charge injection method. This can be attributed to the weakening of the force and displacement of the detection frame by the quadrature force nulling method. Figures 4.26 and 4.27 show the simulation results of V_{QEF} and V_{fQEF} after quadrature force nulling. In a short time, the system reaches the steady state and keeps the steady state without oscillation. Experiments on the circuit of this method will be described in the following sections because of the problem of how to apply the nulling force.

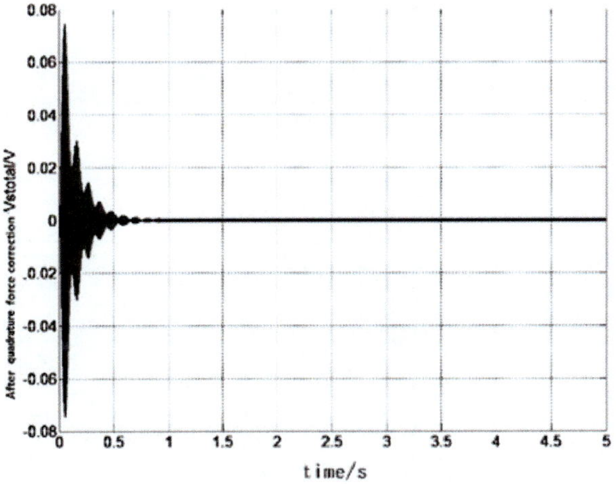

Fig. 4.24 V_{stotal} simulation results after quadrature force correction

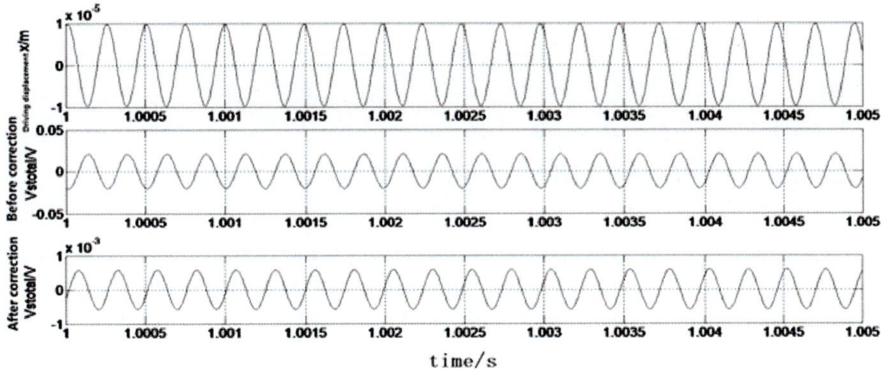

Fig. 4.25 Phase diagram of V_{stotal} before and after quadrature force correction

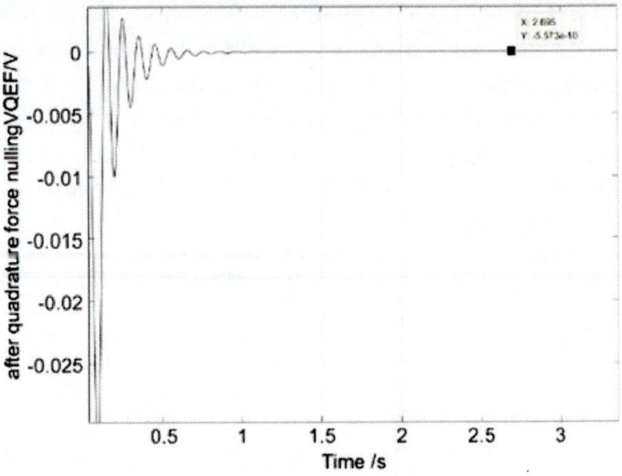

Fig. 4.26 V_{QEF} simulation results after quadrature force correction

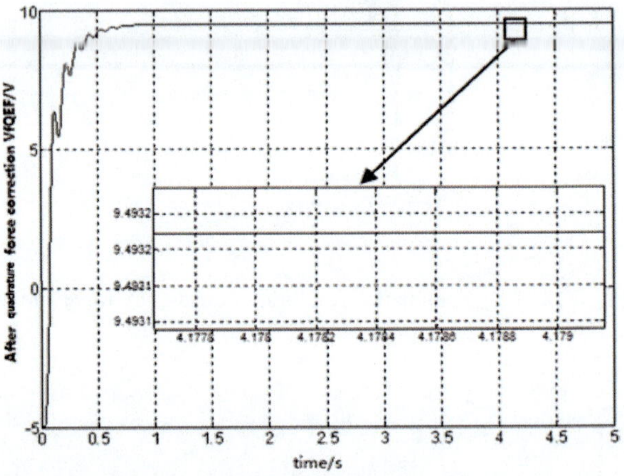

Fig. 4.27 Simulation results of V_{fQEF} after quadrature force correction

4.4.4 *Quadrature Coupling Stiffness Correction Method*

According to the stiffness matrix in formula (4.20), coupling stiffness must be eliminated in order to fundamentally eliminate the quadrature error, that is, $k_{xy} = -k_{qxy}$. At the same time, it can be seen from Fig. 4.7 that the method needs to add an auxiliary nulling mechanism into the structure to generate static negative stiffness, and form a coupling stiffness nulling system with the control circuit [141–144]. Under the control of the external circuit, the quadrature coupling stiffness nulling comb teeth generate the corresponding electrostatic negative stiffness to offset the quadrature stiffness, to achieve the purpose of quadrature correction.

4.4.4.1 Analysis of Quadrature Coupling Stiffness Nulling Structure (Quadrature Nulling Comb Tooth)

The quadrature correction comb has been briefly introduced in Fig. 2.5 in chapter 2, and an enlarged schematic diagram of its structure is shown in Fig. 4.28 [94,141]. The structure of each unit is also enlarged in Fig. 4.28: the length of the overlapping part of the comb teeth is X_{Q0} along the x direction, the length of the comb teeth is unequal, the long distance is λ_{yq0}, and the short distance is Y_{Q0}.This section takes the left mass block as an example to analyze the working principle of comb tooth.

First mark left mass and four overlapping eight fixed electrode capacitance (the top comb space is smaller, the tooth top–bottom spacing is bigger, so ignore the capacitance between the comb tooth top to bottom), the upper left and lower left, upper right and lower electrode wire connected respectively by the substrate, and in the quality between the two groups of electrodes and voltage V_{qkl1} and V_{qkl2}.When the mass block has displacement x and y towards the drivinge axis and the detection axis, the capacitance can be expressed in matrix form:

$$\begin{bmatrix} C_{ql1s} & C_{ql2s} & C_{ql3s} & C_{ql4s} \\ C_{ql1x} & C_{ql2x} & C_{ql3x} & C_{ql4x} \end{bmatrix} = \varepsilon_0 h \begin{bmatrix} \frac{1}{\lambda y_{q0}-y} & \frac{1}{y_{q0}-y} \\ \frac{1}{y_{q0}+y} & \frac{1}{\lambda y_{q0}+y} \end{bmatrix} \begin{bmatrix} x_{q0}-x & 0 & 0 & x_{q0}+x \\ 0 & x_{q0}+x & x_{q0}-x & 0 \end{bmatrix} \tag{4.40}$$

According to the method of capacitance electrostatic force analysis of parallel plate introduced in Sect. 2.4, force matrices of each capacitor applied to the mass block along the driving and sensing shafts are obtained respectively (positive directions of the driving and sensing shafts are positive):

Fig. 4.28 Schematic diagram of quadrature coupling stiffness correction comb teeth

$$\begin{bmatrix} F_{xC_{ql1s}} & F_{xC_{ql2s}} & F_{xC_{ql3s}} & F_{xC_{ql4s}} \\ F_{xC_{ql1x}} & F_{xC_{ql2x}} & F_{xC_{ql3x}} & F_{xC_{ql4x}} \end{bmatrix} = \frac{1}{2}\begin{bmatrix} \frac{\partial C_{ql1s}}{\partial x} & \frac{\partial C_{ql2s}}{\partial x} & \frac{\partial C_{ql3s}}{\partial x} & \frac{\partial C_{ql4s}}{\partial x} \\ \frac{\partial C_{ql1x}}{\partial x} & \frac{\partial C_{ql2x}}{\partial x} & \frac{\partial C_{ql3x}}{\partial x} & \frac{\partial C_{ql4x}}{\partial x} \end{bmatrix} V_{qkl}^2$$
$$= \frac{\varepsilon_0 h}{2}\begin{bmatrix} \frac{1}{\lambda y_{q0}-y} & \frac{1}{y_{q0}-y} \\ \frac{1}{y_{q0}+y} & \frac{1}{\lambda y_{q0}+y} \end{bmatrix}\begin{bmatrix} -V_{qkl1}^2 & 0 & 0 & V_{qkl2}^2 \\ 0 & V_{qkl1}^2 & -V_{qkl2}^2 & 0 \end{bmatrix} \tag{4.41}$$

$$\begin{bmatrix} F_{yC_{ql1s}} & F_{yC_{ql2s}} & F_{yC_{ql3s}} & F_{yC_{ql4s}} \\ F_{yC_{ql1x}} & F_{yC_{ql2x}} & F_{yC_{ql3x}} & F_{yC_{ql4x}} \end{bmatrix} = \frac{1}{2}\begin{bmatrix} \frac{\partial C_{ql1s}}{\partial y} & \frac{\partial C_{ql2s}}{\partial y} & \frac{\partial C_{ql3s}}{\partial y} & \frac{\partial C_{ql4s}}{\partial y} \\ \frac{\partial C_{ql1x}}{\partial y} & \frac{\partial C_{ql2x}}{\partial y} & \frac{\partial C_{ql3x}}{\partial y} & \frac{\partial C_{ql4x}}{\partial y} \end{bmatrix} V_{qkl}^2$$
$$= \frac{\varepsilon_0 h}{2}\begin{bmatrix} \frac{1}{(\lambda y_{q0}-y)^2} & \frac{1}{(y_{q0}-y)^2} \\ \frac{-1}{(y_{q0}+y)^2} & \frac{-1}{(\lambda y_{q0}+y)^2} \end{bmatrix} \times \tag{4.42}$$
$$\begin{bmatrix} (x_{q0}-x)V_{qkl1}^2 & 0 & 0 & (x_{q0}+x)V_{qkl2}^2 \\ 0 & (x_{q0}+x)V_{qkl1}^2 & (x_{q0}-x)V_{qkl2}^2 & 0 \end{bmatrix}$$

In the formula, V_{qkl} is the voltage corresponding to each capacitor. Take the sum of all elements of the above two matrices respectively, then the value is the resultant force in one direction. Considering the number of corrected comb teeth n_q, then:

$$F_{xC_{ql}} = n_q \sum_{i=1}^{4}\left(F_{xC_{qlis}} + F_{xC_{qlix}}\right)$$
$$= \frac{\varepsilon_0 h n_q\left(V_{qkl2}^2 - V_{qkl1}^2\right)}{2}\left(\frac{1}{\lambda y_{q0}-y} - \frac{1}{y_{q0}-y} + \frac{1}{y_{q0}+y} - \frac{1}{\lambda y_{q0}+y}\right) \tag{4.43}$$

$$F_{yC_{ql}} = n_q \sum_{i=1}^{4}\left(F_{yC_{qlis}} + F_{yC_{qlix}}\right)$$
$$= \frac{\varepsilon_0 h n_q}{2}\left(\frac{(x_{q0}+x)V_{qkl2}^2 + (x_{q0}-x)V_{qkl1}^2}{(\lambda y_{q0}-y)^2} + \frac{(x_{q0}-x)V_{qkl2}^2 + (x_{q0}+x)V_{qkl1}^2}{(y_{q0}-y)^2}\right.$$
$$\left. - \frac{(x_{q0}+x)V_{qkl2}^2 + (x_{q0}-x)V_{qkl1}^2}{(y_{q0}+y)^2} - \frac{(x_{q0}-x)V_{qkl2}^2 + (x_{q0}+x)V_{qkl1}^2}{(\lambda y_{q0}+y)^2}\right) \tag{4.44}$$

After taking the partial derivatives of x and y on the left and right sides of the above two equations, the stiffness matrices of these two forces on the driving and detection axes can be obtained:

$$k_q = \begin{bmatrix} k_{qxx} & k_{qxy} \\ k_{qyx} & k_{qyy} \end{bmatrix} = -\frac{n_q \varepsilon_0 h}{y_{q0}^2}\begin{bmatrix} 0 & \left(1-\frac{1}{\lambda^2}\right)\left(V_{qkl1}^2 - V_{qkl2}^2\right) \\ \left(1-\frac{1}{\lambda^2}\right)\left(V_{qkl1}^2 - V_{qkl2}^2\right) & \frac{2x_{q0}}{y_{q0}}\left(1+\frac{1}{\lambda^3}\right)\left(V_{qkl1}^2 + V_{qkl2}^2\right) \end{bmatrix} \tag{4.45}$$

In the above equation, the auxiliary diagonal element is the coupling nulling stiffness, and the two-coupling stiffness are equal. In combination with Eq. (4.5), when $k_{qxy} + k_{xy} = 0$, there is $k_{Qyx} + k_{yx} = 0$, then the coupling stiffness of sensing and driving modes can be corrected simultaneously. After substituting into the correlation equation:

$$\frac{k_x - k_y}{2}\sin 2\beta_{Qx} = \frac{n_q\varepsilon_0 h}{y_{q0}^2}\left(1 - \frac{1}{\lambda^2}\right)\left(V_{qkl1}^2 - V_{qkl2}^2\right) \tag{4.46}$$

In the above formula, β_{Qx} is the machining error angle determined by the machining error. After the structural machining is completed, only the square deviation of voltage at the right end of the above formula can be changed, which is beneficial to the design of the control system. In addition, k_{QYY} in the calibration stiffness matrix may reduce the resonant frequency of the sensing mode, and the influence of k_{QYY} on the sensing mode resonant frequency can be reduced by increasing the calibration comb spacing y_{Q0} and decreasing the overlap length x_{q0}. Figure 4.29 shows local photos of quadrature comb teeth.

In order to simplify the control model and reduce the nonlinearity of the control voltage, and to ensure the nulling effect of the calibration structure when β_{Qx} is positive or negative, the voltage loading mode in this book is set as:

$$V_{qkl1} = V_{qD} + V_{qc} \tag{4.47}$$

$$V_{qkl2} = V_{qD} - V_{qc} \tag{4.48}$$

Combining Eqs. (4.45), (4.47) and (4.48), we can obtain:

$$k_{qxy} = k_{qyx} = k_{qcoup}V_{qD}V_{qc} = -\frac{4n_q\varepsilon_0 h}{y_{q0}^2}\left(1 - \frac{1}{\lambda^2}\right)V_{qD}V_{qc} \tag{4.49}$$

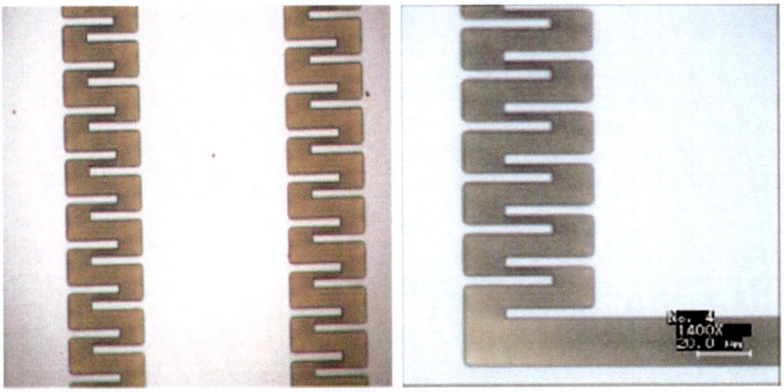

Fig. 4.29 Photos of quadrature stiffness nulling comb teeth

Table 4.3 Parameter coupling stiffness nulling comb design parameters

Parameter	Parameter
Calibration gear unequal pitch ratio λ	2.4
Nulling tooth spacing y_{q0}	4 μm
Nulling tooth overlaps length x_{q0}	20 μm
Corrected fixed voltage V_{qD}	5 V
permittivity of vacuum ε_0	8.85×10^{-12}F/m
Calibrating tooth height h	60 μm
Number of corrective teeth n_q	45

$$k_{qyy} = k_{qsense}\left(V_{qD}^2 + V_{qc}^2\right) = -\frac{4x_{q0}n_q\varepsilon_0 h}{y_{q0}^3}\left(1 + \frac{1}{\lambda^3}\right)\left(V_{qD}^2 + V_{qc}^2\right) \tag{4.50}$$

where, k_{qcoup} and k_{qsense} are structural coupling stiffness coefficient and structural detection stiffness coefficient respectively in Fig. 4.6. Table 4.3 shows the design parameters of the calibration comb. After plugging in the corresponding data in Table 4.3, $k_{qcoup} = -0.0049$ N/m/V^2 and $K_{qsense} = -0.032$ N/m/V^2 can be obtained. According to the data in the table, the influence curve of calibration voltage V_{qc} on machining error angle β_{Qx} and total modal stiffness k_{yy} can be drawn, as shown in Fig. 4.30.

The solid black line is the relationship between V_{qc} and β_{Qx}. The hyperbola shows the relationship between V_{qc} and K_{YY}. As can be seen from the figure, V_{qc} changes in the range of −10 V to +10 V (the actual range of change is related to circuit

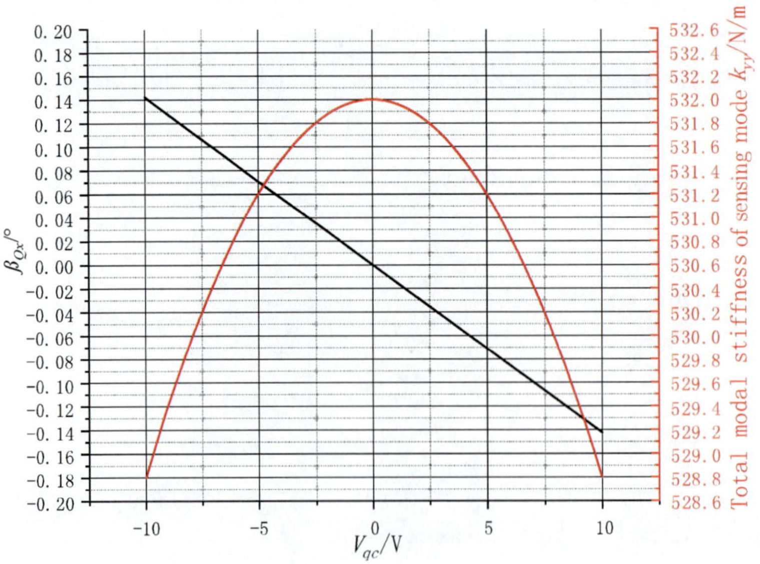

Fig. 4.30 Influence of calibration voltage on machining error angle and detection modal stiffness

design, the figure is only quantitative analysis), and its ability to correct the angle of machining error increases linearly. When it is 10 V, the β_{Qx} that can be corrected is about 0.14° (corresponding to the equivalent input angle rate Ω_{QE} is 176°/s). It proves that the normal structure proposed in this book has high efficiency. At the same time, the influence of normal structure on modal stiffness detection is very weak. When the V_{qc} changes within the range of ±10 V, the change of detected modal stiffness is less than 1%, which will have a certain influence on the frequency difference between driving and detecting mode and change the scale factor value. When the temperature changes, the stiffness of the driving and sensing modes will change accordingly. According to Eq. (4.5), the coupling stiffness will also change accordingly, so will the V_{qc}.

4.4.4.2 System Design of Quadrature Coupling Stiffness Correction Method

Compared with the charge injection method and quadrature force correction method introduced previously, the method described in this section does not have the modulation part of the driving force AC signal, and avoids the secondary influence of the phase error and noise of the modulation signal. Figure 4.31 shows the block diagram of unilateral quality control:

The method to obtain quadrature signals in the figure is the same as the previous method. Referring to the system analysis method in the previous section, the mature AGC closed-loop control method is also used to design the control system. The following equation can be obtained from Fig. 4.1:

$$V_{stotal} = \left(F_{QE} + F_{IP}\right)G_{sV/F} = \left[xk_{yxs} + 2(\Omega_z + \Omega_{IP})A_x m_c \omega_d \sin(\omega_d t)\right]G_{sV/F} \tag{4.51}$$

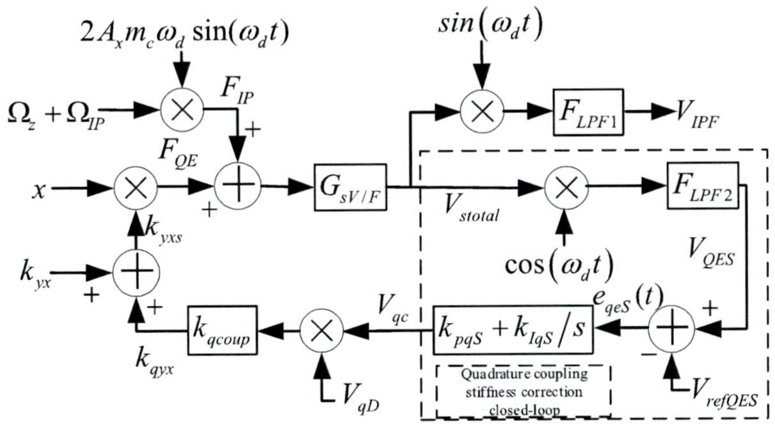

Fig. 4.31 Working principle diagram of quadrature coupling stiffness correction method system

$$k_{yxs} = k_{yx} + k_{qcoup} V_{qD} V_{qc} \tag{4.52}$$

where, $x = A_{x\cos(\omega dt)}$ and $V_{refQES} = 0$. If the action of low-pass filter F_{LPF2} is considered, then:

$$V_{QES} = [V_{stotal}\cos(\omega_d t)]\big|_{F_{LPF2}} = \frac{A_x(k_{yx} + k_{qyx})G_{sV/F}}{2} \tag{4.53}$$

$$V_{qc} = V_{QES}\left(k_{pqS} + \frac{k_{IqS}}{s}\right) \tag{4.54}$$

4.4.4.3 Analysis of System Characteristics by Quadrature Coupling Stiffness Correction Method

It can be seen from Eq. (4.53) that the control signal is only related to the size of the quadrature component. In order to better reflect the static and dynamic characteristics of the system, the coupling stiffness $k_{yx}(t)$ is regarded as a time-related variable as the system input, and $k_{qyx}(t)$ as the system output. The analysis method similar to the previous section is not difficult to obtain:

$$\frac{k_{qyx}(s)}{k_{yx}(s)} = \frac{\frac{A_x}{2}k_{qcoup}V_{qD}F_{LPF2}(s)\left[G_{sV/F}(s + j\omega_d) + G_{sV/F}(s - j\omega_d)\right]\left(k_{pqS} + \frac{k_{IqS}}{s}\right)}{1 - \frac{A_x}{2}k_{qcoup}V_{qD}F_{LPF2}(s)\left[G_{sV/F}(s + j\omega_d) + G_{sV/F}(s - j\omega_d)\right]\left(k_{pqS} + \frac{k_{IqS}}{s}\right)} \tag{4.55}$$

The above formula can be expressed in Fig. 4.32, and when the system is at a static state, $s = 0$, then in the denominator on the right side of Eq. (4.55):

$$1 \ll \frac{A_x}{2}k_{qcoup}V_{qD}F_{LPF2}(s)\left[G_{sV/F}(s + j\omega_d) + G_{sV/F}(s - j\omega_d)\right]\left(k_{pqS} + \frac{k_{IqS}}{s}\right)\Bigg|_{s=0} \tag{4.56}$$

Then:

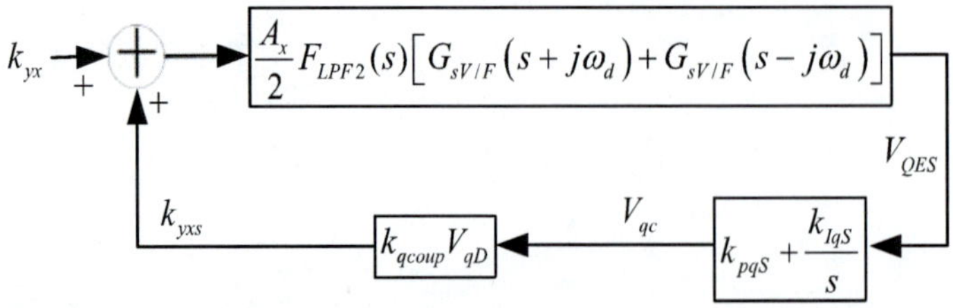

Fig. 4.32 Simplified schematic diagram of the quadrature coupling stiffness correction method

$$k_{qyx} \approx -k_{yx} \tag{4.57}$$

The static negative stiffness can cancel the quadrature stiffness. By substituting the data in Table 4.4 into the transmission function of the control system, it can be obtained that the system's amplitude margin and phase margin are 5 dB and 89° respectively in the case of open loop, as shown in Fig. 4.33. The amplitude margin is determined by the integral link in the PI controller, which also affects the system bandwidth in the closed-loop state.

The zero-pole distribution and Nyquist curve of the system in closed-loop state are shown in Figs. 4.34 and 4.35. As shown in Fig. 4.34, the poles of the system are located in the negative half axis of the real axis, and the Nyquist curve does not contain $(-1, 0)$, it can be determined that the closed-loop state of the system is stable (Fig. 4.36).

Table 4.4 Model parameters of each link of quadrature stiffness correction system

The parameter name	Parament value
Coupling stiffness coefficient of quadrature correction comb tooth k_{qcoup}	-0.0049 N/m/V^2
DC component of comb voltage V_{qD}	5 V
PI controller proportional coefficient k_{pqS}	0.01
PI controller integral coefficient k_{IqS}	2,500,000

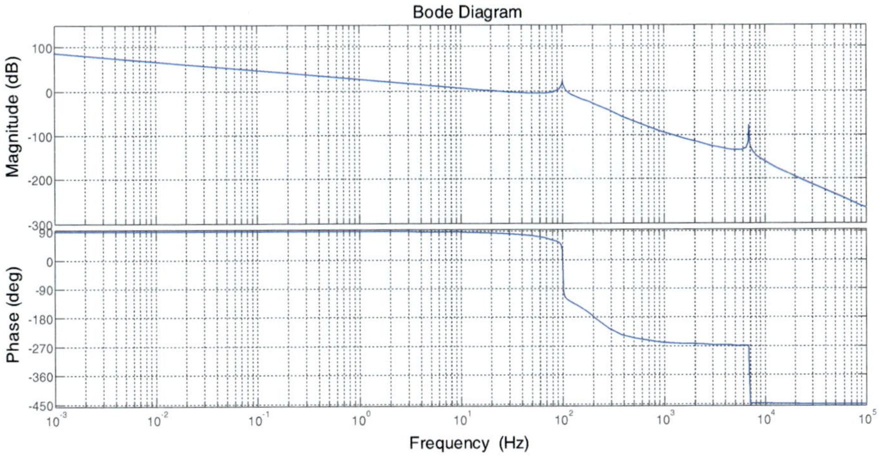

Fig. 4.33 Open-loop Bod diagram of quadrature coupling stiffness correction method

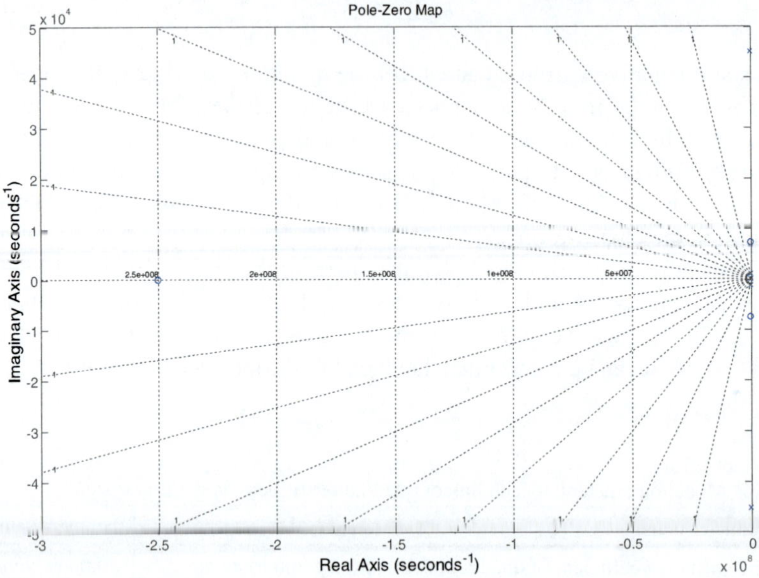

Fig. 4.34 Distribution of zero and pole in closed-loop of quadrature coupling stiffness correction method

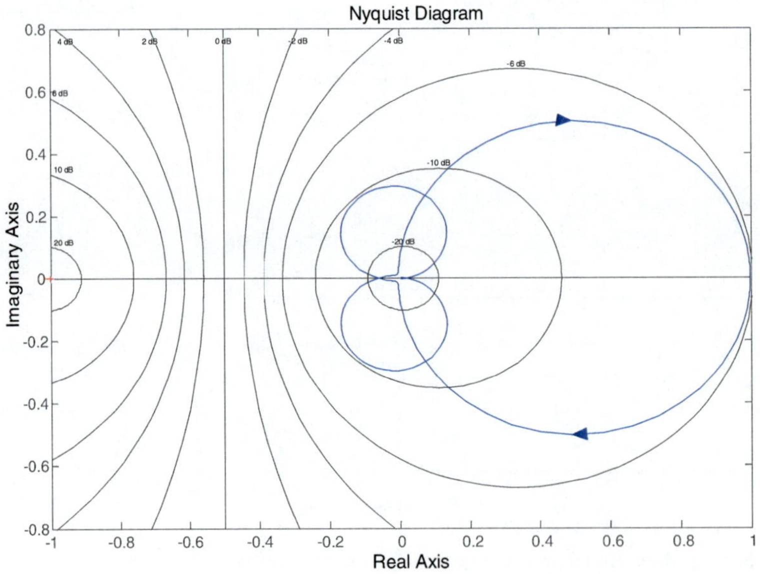

Fig. 4.35 Closed-loop Nyquist curve of quadrature coupling stiffness correction method

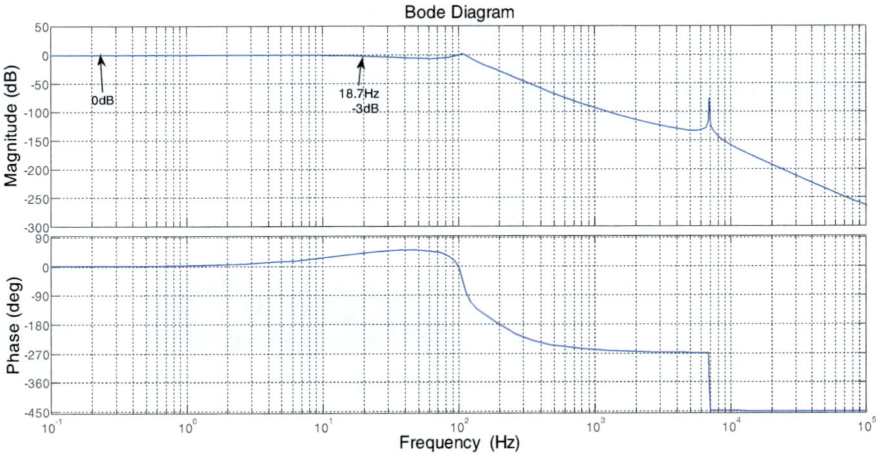

Fig. 4.36 Closed-loop bod diagram of quadrature coupling stiffness correction method

4.4.4.4 Quadrature Coupling Stiffness Correction Method System Time Domain Simulation

According to the control method in the above section, a control loop is built in the Simulink model, and the whole model is simulated after PI control parameters are adjusted. The simulation model is shown in Fig. 4.37:

After the quadrature coupling stiffness was corrected, the output signal V_{stotal} and control signal V_{pc} curves of the gyro detecting modal interface are shown in Figs. 4.38 and 4.39. In a short time, the system reached the steady state, and the V_{qc} signal was about 6.66 V. As shown in the figure, the control signal was very stable without fluctuation, and the performance of the control system was good. In order

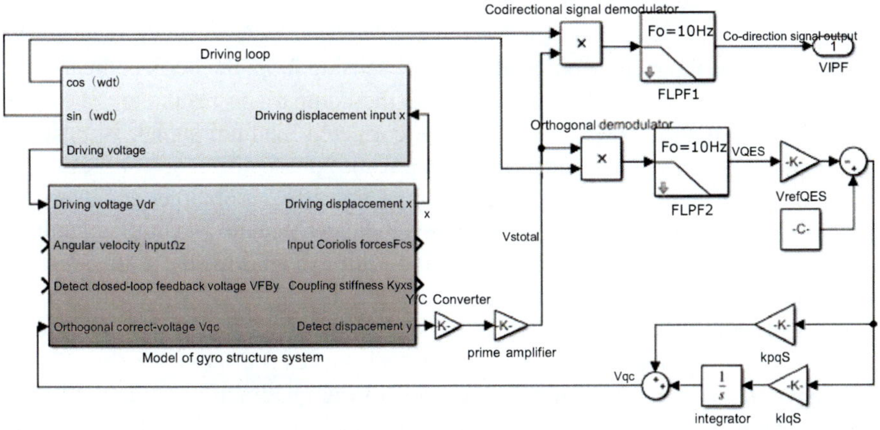

Fig. 4.37 System simulation diagram of quadrature coupling stiffness correction method

Fig. 4.38 V_{stotal} signal simulation curve after quadrature coupling stiffness correction

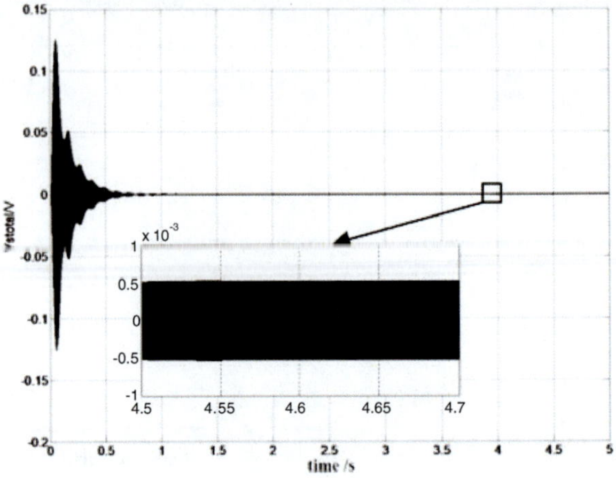

Fig. 4.39 V_{qc} signal simulation curve after quadrature coupling stiffness correction

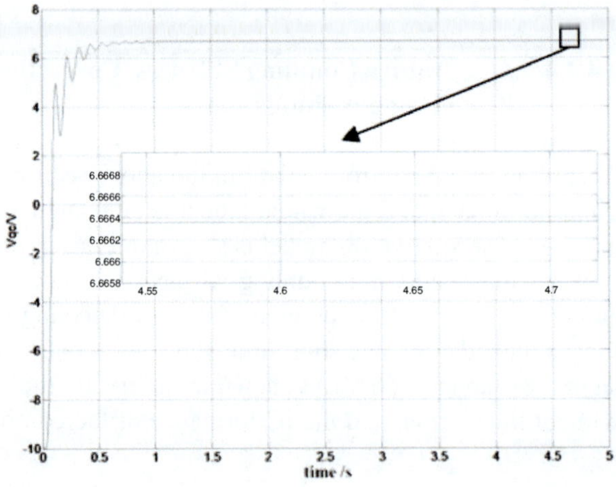

to better observe the signal composition in the sensing loop, the book compares the driving displacement signal with V_{stotal}, and the comparison results are shown in Fig. 4.40. The figure shows that V_{stotal}, the corrected medium signal, is basically only contained in the Coriolis identical signal whose phase difference of the driving displacement signal is 90°. The quadrature correction effect is obvious, and its amplitude is consistent with that in Fig. 4.19, which indicates that the two nulling methods can achieve the same nulling effect under the condition of ignoring the influence of the phase frequency of the modulation signal. However, due to the influence of electronic component error and temperature change and other unsatisfactory factors, the modulation signal will produce error and noise, which makes it difficult for this method to achieve the state in the simulation model, so the quadrature coupling stiffness correction method has more obvious advantages. Further experimental comparisons of the two methods are detailed in the following sections.

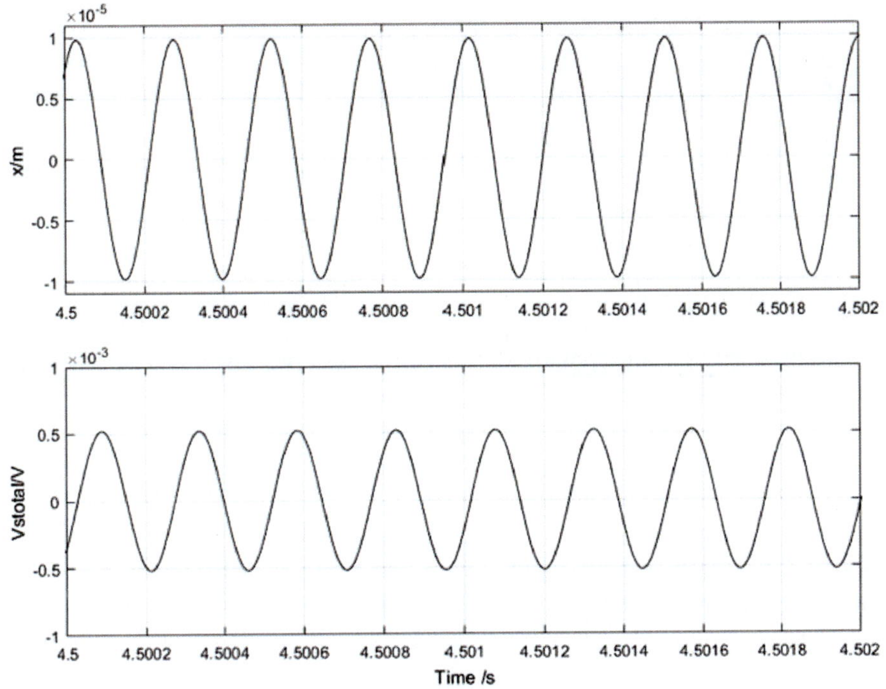

Fig. 4.40 Phase comparison curve of simulation results of V_{stotal} signal and driving displacement signal x after quadrature coupling stiffness correction

In order to further verify the dynamic performance of the system, step input is used as the correction DC signal V_{qD} in this book, and the closed-loop loop is enabled to work at 2 s, and the coupling stiffness $k_{yxs} = k_{yx} + k_{Qyx}$ is observed. The results are shown in Fig. 4.41. In the figure, the k_{yXS} curves of the system without and after nulling are enlarged. Before nulling, the value is k_{yx}, while after quadrature correction, the value is about 3.29×10^{-4} N/m, which is more than 400 times smaller, greatly weakened the influence of coupling stiffness.

For the above three different quadrature nulling methods, their feasibility and reliability have been proved in system simulation. However, it can be seen from Fig. 4.7 that the quadrature coupling stiffness correction method targets at the root of quadrature error (quadrature coupling stiffness). The quadrature force nulling method corresponds to the quadrature force (the product of the quadrature coupling stiffness and the driving displacement), which is inevitably affected by the driving displacement factor, and it does not eliminate the root of the quadrature error. The charge injection method only deals with the signal, and the quadrature motion of the structure still exists. Therefore, from the theoretical and logical analysis, the stiffness nulling method is the best, the quadrature force nulling method is the second, and the charge injection method is the worst. In the following section, the respective

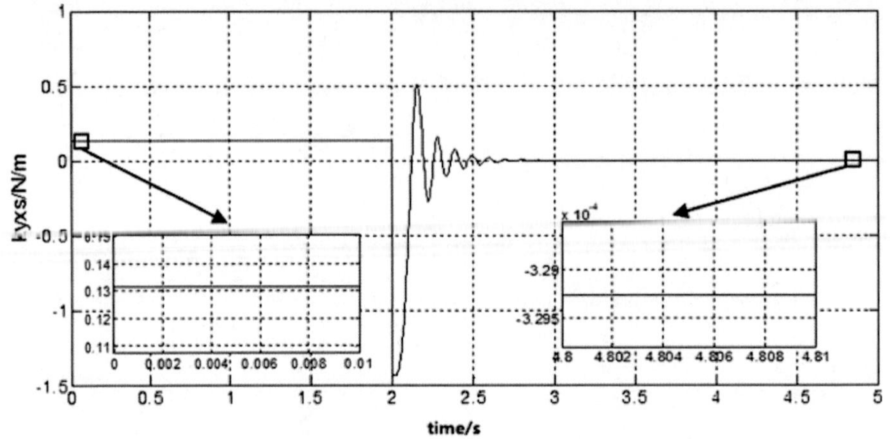

Fig. 4.41 Comparison of total stiffness before and after quadrature coupling stiffness correction

methods will be further verified and compared in actual circuits in order to get the most suitable quadrature nulling method in this book.

4.5 Optimization of Quadrature Correction Method for Dual-Mass Linear Vibration Silicon-Based MEMS Gyroscope

This section introduces the specific realization mode and optimal nulling method of quadrature nulling for dual-mass linear vibration silicon-based MEMS gyroscope, analyzes the characteristics of two different modes of action: single and whole correction of mass block, and selects a more effective mode of action. Then, on the basis of corresponding correction methods, charge injection method, quadrature force correction method and coupling stiffness correction method were used to conduct calibration experiments on the same gyro, and the optimal scheme was selected by analyzing the correction results, which provided the basis for the next test.

When a dual-mass linear vibration silicon-based MEMS gyroscope works, its two masses vibrating at the same frequency, so two masses can be equivalent to a solid block of gyroscope (its sensing mode resonance frequency is the anti-phase mode), which brings two quadrature correction methods: two masses separate correction and together correction.

4.5.1 Dual-Mass Quadrature Error Together Correction

The quadrature correction method involved in this section is the together correction of two mass blocks, which is characterized by applying the same control signal to the two mass blocks at the same time. For example, $V_{qkl1} = V_{qkr2}$, $V_{qkl2} = V_{qkr1}$ are shown in Fig. 4.28. The advantage of this method is that it can save one correction voltage loop, reduce the complexity of the whole circuit system, reduce the power consumption, and theoretically good for reducing the drift of the gyroscope (mostly due to the influence of heat generated after the circuit is started).

It can be seen from the analysis of the above sections that the quadrature nulling method can completely eliminate the quadrature force or stiffness of the structure. Therefore, one of the most critical links of the system is the detection of quadrature signals. By means of the equivalent model in Fig. 4.1, a certain state exists in the gyro structure during movement, as shown in Fig. 4.42:

In the figure, β_{QxL} and β_{QxR} are the included angles of quadrature errors of the left and right mass blocks respectively. V_{sL} and V_{sR} are output signals of left and right mass preamplifiers respectively. Left—right drive and detection frame to the direction of motion. It can be seen that the V_{stotal} signal is the superposition of the two mass blocks' respective output differential signals, which can greatly eliminate the common mode error. For the convenience of analysis, k_x and k_y are split into k_{xL} and k_{xR}, k_{yL} and k_{yR}, which are the stiffness of the left and right mass blocks along the x and y directions respectively. According to Eq. (4.5), the quadrature signal components in V_{sL} and V_{sR} can be written:

$$V_{sLQUER} = x\frac{k_{yL} - k_{xL}}{2}\sin 2\beta_{QxL} G_{sLV/F} \tag{4.58}$$

$$V_{sRQUER} = x\frac{k_{yR} - k_{xR}}{2}\sin 2\beta_{QxR} G_{sRV/F} \tag{4.59}$$

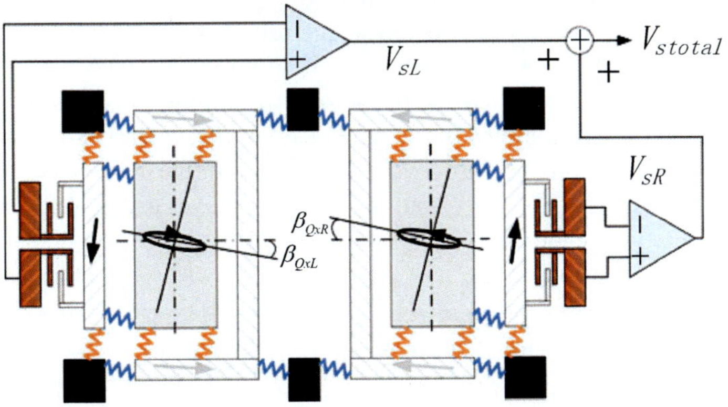

Fig. 4.42 Interface diagram of differential detection for dual-mass silicon micromechanical gyroscope

where $G_{sLV/F}$ and $G_{sRV/F}$ are respectively the product of gyro structure conversion function and interface circuit conversion coefficient composed of left and right mass blocks. If $G_{sLV/F} = G_{sRV/F}$ and $k_{xL} = k_{xR}$ and $k_{yL} = k_{yR}$ are approximately considered, quadrature signal components in V_{stotal} can be expressed according to the above two expressions as follows:

$$V_{stotalQUER} = \frac{xG_{sLV/F}(k_{yL} - k_{xL})}{2}\left[\sin 2\beta_{QxL} + \sin 2\beta_{QxR}\right] \tag{4.60}$$

As the controlled signal, V_{stotal} is sent into the demodulator and the quadrature nulling controller. Under the action of the quadrature nulling feedback control quantity, $V_{stotalQUER} = 0$. At the same time, considering that the feedback quantity has the same effect on the left and right mass blocks, the following situations must occur (usually $-90° < 2\beta_{QxL} < 90°$ and $-90° < 2\beta_{QxR} < 90°$):

(1) When $\beta_{QxL} = \beta_{QxR}$, quadrature nulling feedback can eliminate the quadrature signals of left and right mass blocks simultaneously.
(2) When β_{QxL} and β_{QxR} are both positive and $\beta_{QxL} > \beta_{QxR}$, when the control system is stable, there will be:

$$\beta_{QxLs} = -\beta_{QxRs} = \frac{\beta_{QxL} - \beta_{QxR}}{2} \tag{4.61}$$

β_{QxLs} and β_{QxRs} are the quadrature error angles after the system is stabilized.

(3) When β_{QxL} and β_{QxR} are both negtive and $\beta_{QxL} > \beta_{QxR}$, when the control system is stable, there will be:

$$\beta_{QxRs} = -\beta_{QxLs} = \frac{\beta_{QxR} - \beta_{QxL}}{2} \tag{4.62}$$

In conclusion, only when $G_{sLV/F} = G_{sRV/F}$, $k_{xL} = k_{xR}$, $k_{yL} = k_{yR}$ and $\beta_{QxL} = \beta_{QxR}$, the global nulling method can eliminate the quadrature motion. However, the parameters of the left and right mass blocks are usually different, so in most cases, the overall nulling of the two mass blocks can only greatly reduce the quadrature motion, but not eradicate it. In view of the overall nulling mode of the two mass blocks, this section takes the coupling stiffness nulling method as an example to carry out the overall nulling of the two mass blocks. In the case of no quadrature nulling, the output signal of gyro front discharge is shown in Fig. 4.43, where the first line of curve is the driving displacement signal, the second and third lines are V_{sL} and V_{sR}, and the fourth line is V_{stotal}. It can be seen from the figure that the peak-to-peak values of V_{sL} and V_{sR} are 150 mV and 300 mV respectively, and both are in reverse phase with the driving displacement signal.

Figure 4.44 is the signal curve of each point after the overall quadrature nulling. It can be seen from the figure that V_{sL} and V_{sR} attenuated significantly compared with before nulling, with peak-to-peak values falling below 40 mV, and the phases were opposite. Although there were no obvious quadrature signal characteristics in

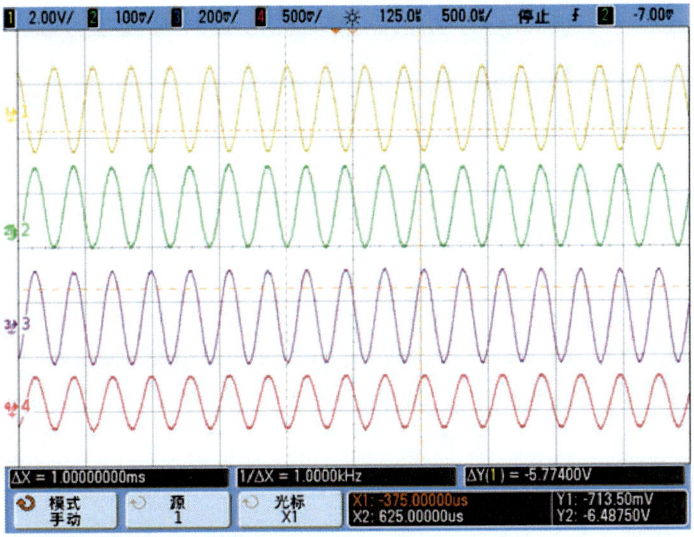

Fig. 4.43 Quadrature signal diagram in the detection channel before quadrature nulling

V_{stotal}, However, from the phase relationship between V_{sL} and V_{sR} and the driving displacement signal, we can see that the quadrature signal is still dominant in their signal composition, that is, the method does not weaken the quadrature signal to the maximum extent.

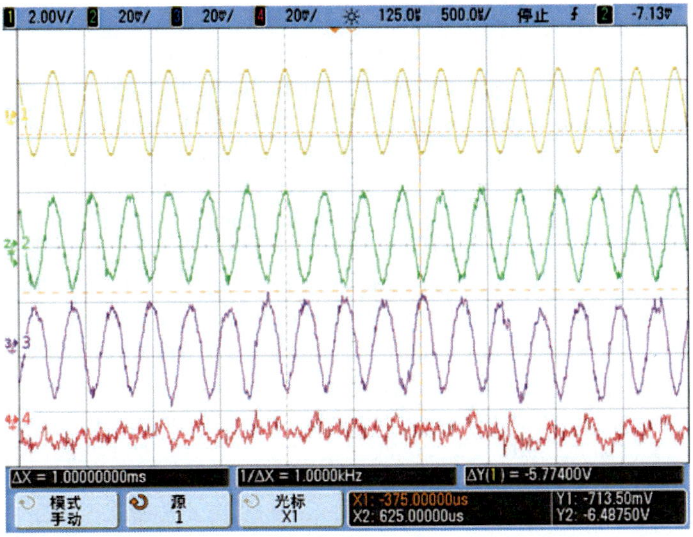

Fig. 4.44 Quadrature signals in sensing loop channel after two masses together correction

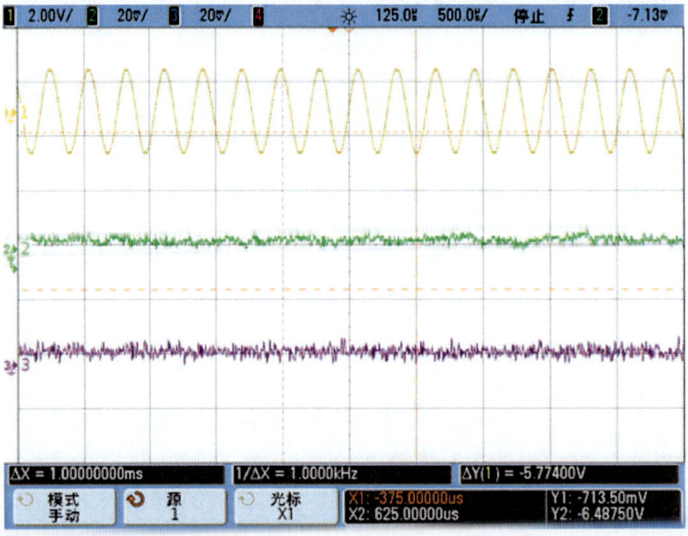

Fig. 4.45 Separate correction experiment of two mass blocks by quadrature coupling stiffness nulling method

4.5.2 Dual-Mass Quadrature Error Separate Correction

The quadrature error separate correction method of two mass blocks is to independently design the closed-loop loop for left and right mass blocks respectively, and the correction signals are not connected to each other. In Fig. 4.45, V_{sL} (curve in the second line) and V_{sR} (curve in the third line) are input to their respective quadrature nulling closed-loop as the controlled signals of the left and right mass blocks respectively. The control objects are Eq. (4.53) and (4.54). After the system is stabilized, $V_{sLQUER} = 0$ and $V_{sRQUER} = 0$, then the corresponding $\beta_{QxLs} = \beta_{QxRs} = 0$, and the quadrature error angles of the two mass blocks are completely corrected. Compared with the whole nulling method in the previous section, the independent nulling method (as long as the quadrature quantity is in a controllable range) can better achieve the purpose of eliminating the quadrature motion. The three methods described in this chapter were used to independently correct the two mass blocks, and their respective experimental results were shown in the following three Figs. 4.45, 4.46, 4.47 (the first line of curve was the driving displacement signal, and the second and third lines were V_{sL} and V_{sR}). It can be seen from the figure that V_{sL} and V_{sR} have no obvious quadrature signal characteristics, and the quadrature motion of the two mass blocks is well corrected.

As can be seen from the experimental results of the above three methods, although the quadrature signals can be eliminated by the three methods, the nulling effect and noise level of the two mass blocks are not consistent due to the factors of gyro individual and control loop (V_{sL} noise is significantly smaller than V_{sR} noise). Meanwhile, due to different control objects, the effect results of the three methods are

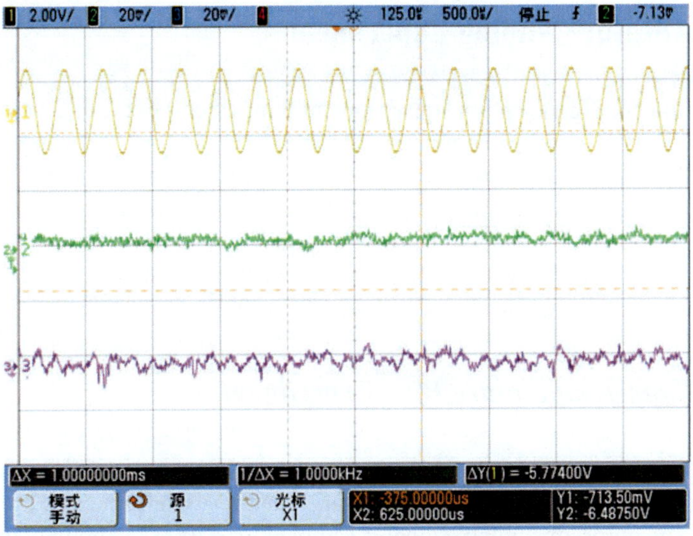

Fig. 4.46 Separate correction experiment of two mass blocks by quadrature force nulling method

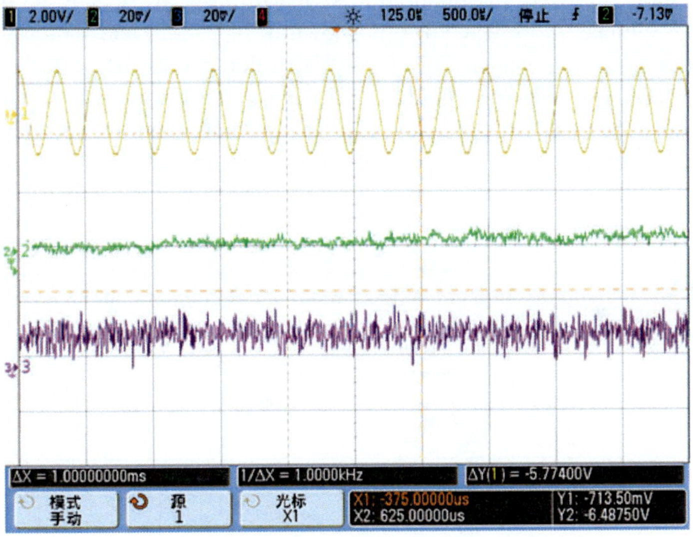

Fig. 4.47 Separate correction experiment of two mass blocks by charge injection method

different to some extent. As can be seen from the above three figures, the quadrature coupling stiffness nulling method is better than the quadrature force nulling method than the charge injection method in terms of noise control effect, which is consistent with the analysis results in the above section. Quadrature coupling stiffness method and two mass block independent nulling method are selected as the optimal scheme to carry out in-depth practical test.

4.6 Quadrature Nulling Experiment

Under the influence of the optimal quadrature nulling scheme of the dual-mass silicon micromechanical gyroscope summarized in the above section, the normal temperature and full temperature characteristics of the gyroscope output signal under the quadrature open-loop and closed-loop conditions were tested, and the output characteristics of the gyroscope under the quadrature open-loop and closed-loop conditions were compared to verify the theoretical research carried out above.

4.6.1 Room Temperature Bias Experiment

When the sensing loop is open and quadrature correction loop is open, the amplitude of quadrature signal in the sensing loop is large, and its waveform can be referred to Fig. 4.43. The gyroscope output signal is sampled for 4800 s (sampling rate: 1 Hz), and the state curve is shown in Fig. 4.48. Its scale factor is −12.5496 mV/°/s, so the zero-bias stability of the state can be calculated as 866°/h. After the quadrature correction loop is closed, the quadrature signal in the gyro sensing loop channel is basically eliminated, its waveform is shown in Fig. 4.45, and the output signal curve of the gyro is shown in Fig. 4.49. With reference to the scale factor of −10.0253 mV/°/s, the zero-bias stability of the gyro after the quadrature nulling can be calculated to be 22.67°/h. As can be seen, the quadrature nulling loop can restrain the drift characteristic of gyro zero offset output and optimize the gyro zero offset performance. More test results are detailed and summarized in Chap. 6.

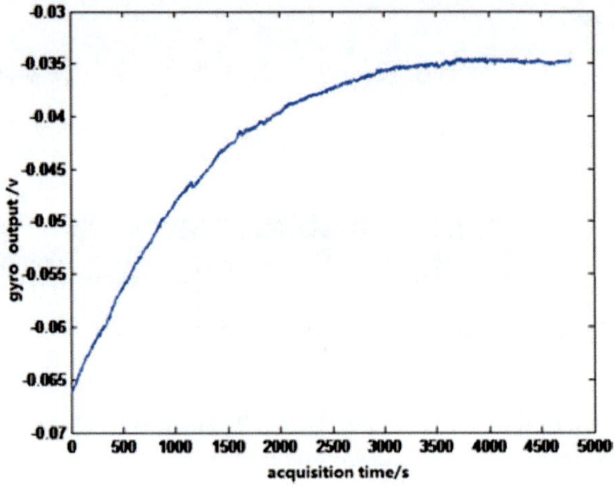

Fig. 4.48 Gyro sensing open-loop output curve before quadrature correction

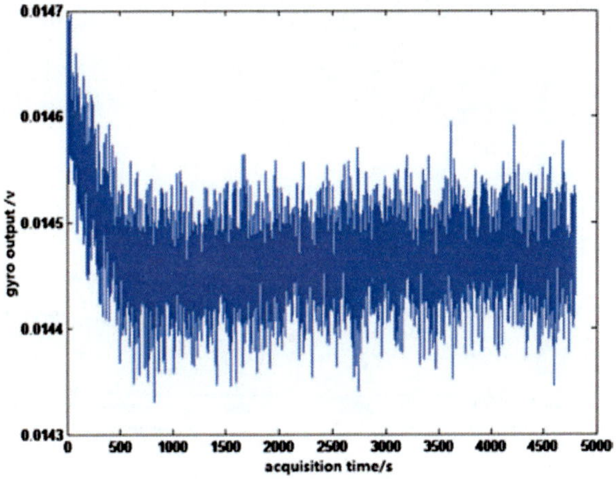

Fig. 4.49 Output curve of gyro sensing open-loop after quadrature correction

4.6.2 Full Temperature Bias Experiment

The V_{qc} control quantities of the left and right mass blocks in Fig. 4.31 were monitored at full temperature, and the temperature setting during the test was the same as Sect. 3.4.4. Curves are shown in Figs. 4.50 and 4.51. This control quantity directly reflects the amplitude of quadrature signals in the detection channel. The above curves prove that the controller can correct the quadrature coupling stiffness well in the full temperature range, with good system stability and reliability, and verify the correctness of the quadrature optimal nulling scheme of the dual-mass silicon micromechanical gyroscope proposed in this chapter.

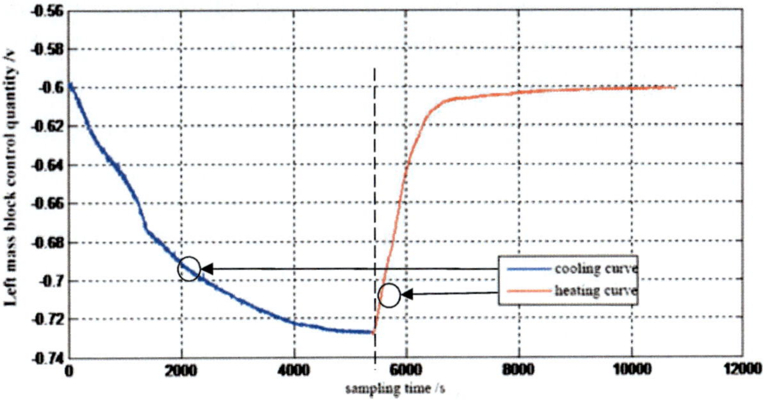

Fig. 4.50 Full temperature curve of the left mass block control results

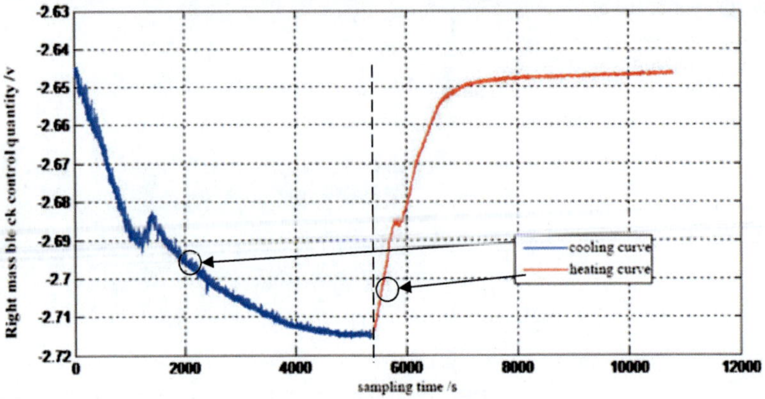

Fig. 4.51 Full temperature curve of the right mass block control results

4.7 Chapter Summary

This chapter first analyzes the structure error of silicon micromechanical gyroscope, which is mainly caused by the machining error of silicon structure, mainly in the form of structural unequal elasticity and damping asymmetry, and makes qualitative and quantitative analysis of the principle of these unsatisfactory factors. Secondly, the composition of the output signal of the gyroscope is analyzed, and the quadrature error signal is defined according to the phase characteristics of these signals. Thirdly, the quadrature module is added to the gyro structure model proposed in the previous chapter, and three different quadrature correction methods are developed through the study of the mode of action of the quadrature error, which are the charge injection method for the quadrature signal, the correction method for the quadrature force and the correction method for the quadrature coupling stiffness. Then, based on the mature AGC closed-loop control technology, the above three quadrature nulling models are built in the system simulation model and analyzed and simulated. Then, the application method of quadrature nulling for dual-mass silicon micromechanical gyroscope is introduced, and it is proved that the effect of separate correction of two mass blocks is better than the together correction. Then, the different quadrature correction methods are realized on the circuit, and the experiment proves that the quadrature coupling stiffness nulling method is better than the other two methods, and the quadrature signal is completely eliminated, which proves the results of theoretical analysis. Finally, experiments at room temperature before and after quadrature correction prove that the quadrature nulling can effectively improve the drift trend of gyro, and the zero-bias stability of gyro is improved from 866 to 22.67°/h in open loop state. At the same time, experiments at full temperature state also show that the quadrature control loop can work stably and reliably.

Chapter 5
Dual-Mass Linear Vibration Silicon-Based MEMS Gyroscope Sensing Closed Loop and Frequency Tuning Technology

5.1 Chapter Introduction

For the dual-mass linear vibration silicon-based MEMS gyroscope, the angular velocity information can be obtained from the sensing mode (it also can be called detection mode or detecting mode) loop under the premise that the driving loop works stably. The detection sensing loop is usually divided into open loop mode and closed loop mode. In the open loop state, the sensing mode displacement amplitude is proportional to the input angular rate, and the gyro output is the sensing mode displacement signal. The closed loop is a sensing mode force feedback control system added on the basis of the open loop. The output of the system is the feedback control force signal, and under this condition, the sensing mode displacement is almost zero, so the closed loop has better dynamic characteristics. In addition, the bandwidth characteristics of the sensing open loop are determined by the driving and sensing working frequency differences. A small frequency difference is helpful to improve the mechanical sensitivity and static characteristics of the gyro, but will severely limit the bandwidth. A larger frequency difference can provide a wider bandwidth but reduce the mechanical sensitivity and the signal-to-noise ratio of the gyro output signal. The sensing loop can expand the bandwidth by optimizing the control system under the premise of small frequency difference. Therefore, the sensing loop makes the gyro have high static performance such as high mechanical sensitivity and high signal-to-noise ratio with small frequency difference and large dynamic characteristics such as bandwidth.

Due to the influence of machining error, it is difficult to unify the machined mechanical parameters of individual gyro with the same design parameters, especially the gyro frequency difference parameters, which directly limits the universality of the closed-loop controller. The frequency difference of driving and sensing modes can be automatically adjusted to a set value by frequency tuning technology, which is of great help to improve the mechanical sensitivity of gyro and enhance the generality of closed-loop controller.

H. Cao, *Dual-Mass Linear Vibration Silicon-Based MEMS Gyroscope*,
https://doi.org/10.1007/978-981-19-9247-6_5

5.2 Sensing Mode Force Rebalancing Combs Stimulation Method

Normally, the bandwidth of the gyroscope and the indices such as scale factor need on the calibration of angular vibration table and turntable, and due to reasons, such as mechanical vibration, the angle of the vibration amplitude and frequency of the vibration table provides constrained, make it cannot provide a precise frequency angular vibration environment, so you need to test modal frequency response through other methods. In this section, the equivalent Coriolis force is generated by applying an electrostatic signal to sensing mode force-feedback (or can be called rebalancing) comb to test the dynamic performance of the modes.

5.2.1 Dual-Mass Linear Vibration Silicon-Based MEMS Gyroscope Sensing Mode Force Rebalancing Combs

Similar to the study on electrostatic driving comb in Chap. 2, the sliding-mode comb based on push–pull method is also adopted in the force-feedback comb introduced in this section, because the pressure film comb will generate additional negative electrostatic stiffness affecting the mode's natural frequency. Figure 2.5 shows the location diagram, and Fig. 5.1 shows the partial photos.

Since the comb teeth in the structure have the same size, the electrostatic force of the detection force feedback teeth in the y direction can be calculated according to Eq. (2.43) (the above figure only shows the feedback teeth on one side of a single mass block, and there is another set of feedback teeth on its upper side that is axisymmetric with it):

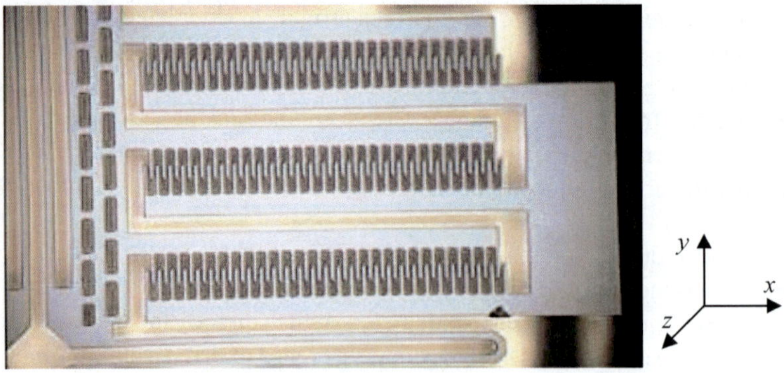

Fig. 5.1 Partial photos of sensing mode force feedback combs

$$F_{yf} = \frac{4n_f \varepsilon_0 h V_{fdc} V_{fac} \sin(\omega_f t)}{d_0} = K_{FBy} V_{fac} \sin(\omega_f t) \tag{5.1}$$

where, $n_f = 86$ is the number of one side of the comb with force feedback. V_{fdc} and V_{fac} are the amplitudes of the DC and AC components of the feedback electrostatic force. ω_f is the angular frequency for detecting the ac component of the feedback electrostatic force. According to Eq. (2.22), the frequency of Coriolis force signal is equal to the driving frequency of the driving mode, that is, $\omega_f = \omega_d$, and the Coriolis force signal is in phase with the driving force signal. If the V_{fdc} term in the fixed form (5.1) has $F_{yf} \propto V_{fac}$, the feedback force can be controlled by controlling the sensing mode feedback AC volume. After calculation, K_{FBy} can be obtained as 1.487 $\times$ 10^{-7} N/V.

5.2.2 Sensing Mode Force Rebalancing Combs Stimulation Method

Literature points out, the Coriolis force can be simulated by apply driving mode AC signal through sensing mode force rebalancing combs [145], but the sensitivity of the method at the expense of the detection signal (as part of the comb is used as the force feedback comb), at the same time also can't form testing differential output, result in output signal electric coupling is serious. The method of detecting force feedback comb tooth excitation proposed in this book is to use force feedback comb tooth to generate electrostatic signal with the same frequency and phase as Coriolis force to observe and test the dynamic performance of mode to Coriolis force. This method can not only keep the sensitivity of the detection signal unchanged, but also test the bandwidth of the detection loop and the dynamic performance of the feedback controller through the method of signal superposition. Figure 5.2 shows the working principle.

In the figure, the dashed line box is the detection closed loop, which does not exist in the detection open-loop state. V_{Oopen} and V_{Oclose} are the output signals of the gyroscope in the open loop and closed loop states respectively. V_{ftest} is an external signal provided by an external signal generator. F_{LPFf} is an output stage low pass filter for the sensing closed loop. Under the condition of sensing mode open loop,

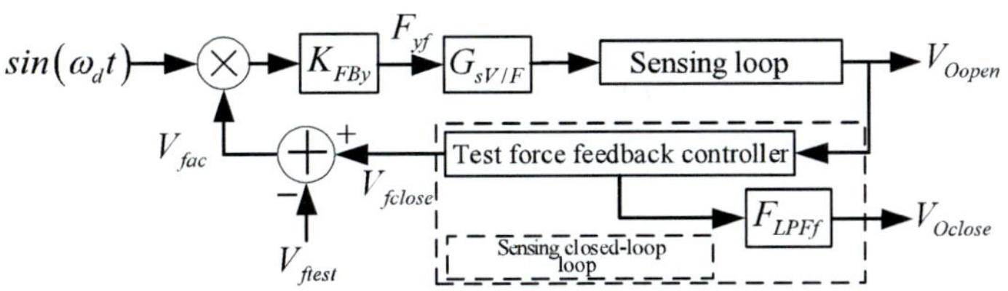

Fig. 5.2 Working principle diagram of sensing mode force feedback combs excitation method

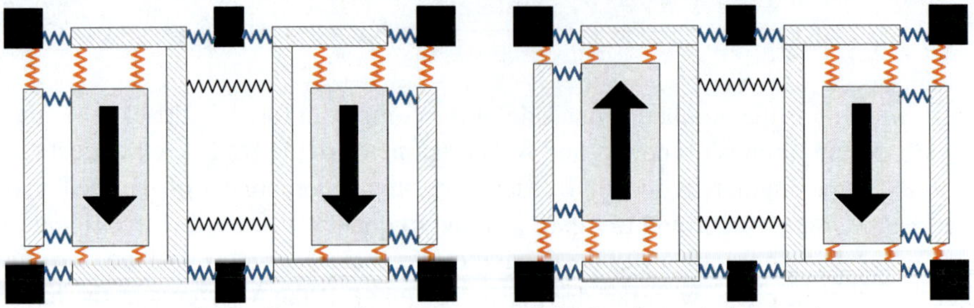

(a) Sensing in-phase mode (b) Sensing anti-phase mode

Fig. 5.3 Schematic diagram of detecting in-phase and anti-phase modal motion

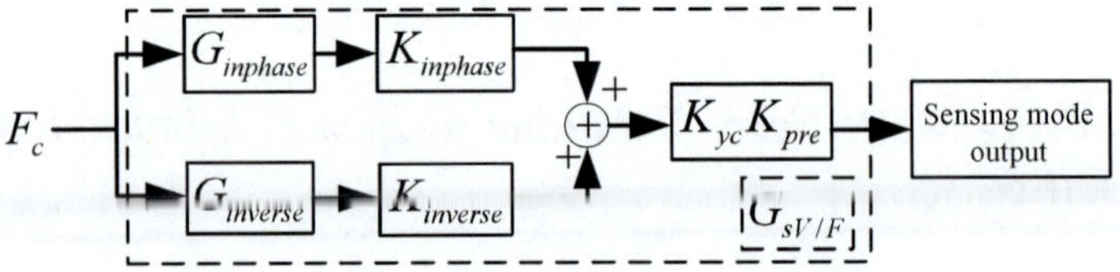

Fig. 5.4 Actual sensing mode system transfer diagram of dual-mass linear vibration silicon-based MEMS gyroscope

set the input rate of the turntable $\Omega_{Z100°/s} = 100°/s$, then according to Eqs. (2.22) and (2.28), the corresponding Coriolis force of the input angle rate can be obtained as shown in Eq. (5.2), and the corresponding output is $V_{Oopen100°/s}$.

$$F_{c100°/s} = 2m_c\Omega_{z100°/s}A_x\omega_d\sin(\omega_d t) \tag{5.2}$$

Maintain the circuit unchanged and make the gyro in a static state. $V_{ftest} = V_{ftest100°/s}$ is reasonably configured to make the output value of the gyro $V_{Oopen100°/s}$, then the electrostatic force $F_{yf100°/s}$ on the comb teeth fed back by the detection force should be equal to $F_{c100°/s}$. The combined Eqs. (5.1) and (5.2) are:

$$K_{FBy}V_{ftest100°/s} = 2m_c\Omega_{z100°/s}A_x\omega_d \tag{5.3}$$

The experimental test data of K_{FBy} is 1.518×10^{-7} N/V when related parameters are substituted, which is nearly equal to the theoretical calculation value in the above section, verifying the accuracy of this parameter. It provides a reliable parameter basis for establishing detection mode and dynamic performance of testing system by force feedback comb excitation method.

5.3 Dual-Mass Linear Vibration Silicon-Based MEMS Gyroscope Sensing Mode Model

It can be seen from the previous chapters that, under working conditions, the detection mode of dual-mass linear vibration silicon-based MEMS gyroscope can convert the input angular rate into Coriolis acceleration and transmit the displacement information of the detection comb tooth to the detection interface. Therefore, the establishment of the sensing mode model and the design of the detection loop are related to the performance and accuracy of the whole gyroscope.

5.3.1 Dual-Mass Silicon Based MEMS Gyroscope Sensing In-phase and Anti-phase Modes Superposition Principle

The two-mass gyro structure is much more complex than the two-single-mass gyro structure, which involves the coupling between the two mass blocks and the superposition of in-phase and anti-phase modes in the field of mechanical vibration [146]. The dual-mass structure adopted in this book has two modes of in-phase and anti-phase detection, but it is generally considered that the anti-phase detection mode is the working mode, as shown in the Fig. 5.3.

In practice, the detection of in-phase mode and anti-phase mode work together because of the superposition of linear vibration modes. In order to obtain more ideal bandwidth and in-band flatness, the frequency difference relationship between different modes should be optimized during the design process. The actual system expression of gyro detection mode is shown in the Fig. 5.4.

In the figure above, $K_{inphase}$ and $K_{inverse}$ are gain multiples of in-phase and inverse-phase modal signals detected respectively, which are mainly determined by signal extraction method and gain and can be obtained through experimental tests. $G_{inphase}$ and $G_{inverse}$ are transfer functions for detecting in-phase mode and anti-phase mode respectively, which are expressed by the following formula:

$$G_{inphase} = \frac{1}{m_y} \frac{1}{s^2 + \frac{\omega_{y1}}{Q_{y1}} s + \omega_{y1}^2}; \quad G_{inverse} = \frac{1}{m_y} \frac{1}{s^2 + \frac{\omega_{y2}}{Q_{y2}} s + \omega_{y2}^2}$$

.

Wherein, ω_{y1}, ω_{y2}, Q_{y1} and Q_{y2} are the natural frequencies and quality factors of sensing in-phase and anti-phase (inverse-phase) modes.

5.3.2 Dual-Mass Linear Vibration Silicon-Based MEMS Gyroscope Sensing Mode Model Establishment and Verification

In order to verify the theory of detecting in-phase and inverse-phase mode superposition proposed in the above section, this section adopts the force feedback comb excitation method to conduct dynamic frequency sweep analysis on the gyro prototype to verify the correctness of the theoretical analysis in the above section, and quantifies the values of $K_{inphase}$ and $K_{inverse}$. In Fig. 5.2, the V_{ftest} point is loaded to test the fixed-amplitude AC voltage, and a series of processes such as pre-amplification, demodulation and low-pass filtering are carried out on the sensing mode output to obtain the actual displacement information of the detected mode. Due to the limitation of the low-pass filter, this section only tests the dynamic characteristics of the low frequency band of the detection mode (below 200 Hz, including the normal working bandwidth band of the gyro). Table 5.1 shows the parameters of sample gyro GY-027 in the open-loop detection and the closed-loop state of quadrature nulling system. Figure 5.5 shows the bod diagram drawn from the experimental data and the simulation diagram of the model derived from the above section.

In the figure, the smooth solid line is the model simulation diagram, and the dotted line with marked points is the actual test point and its connecting line. It can be seen from the figure that the system has two pairs of conjugate poles at A and C, respectively, at the point where the difference between the natural frequency of inverting and in-phase modes and the driving natural frequency is detected (A and C, respectively 3488.9–3464.1 = 24.8 Hz and 3488.9–3360.1 = 128.8 Hz).According to the phase variation, it can be judged that there is a pair of conjugate zeros at B ($\Delta\omega_B$). At the same time, the amplitude of point A is about 33 times that of point C, that is, $K_{inverse} \approx 33K_{inphase}$ (only for GY-027 gyro, there should be differences in the corresponding proportion of different gyro structures).The experiments above prove that the actual detection mode is indeed composed of the superposition of in-phase and antiphase modes, but because of the selective action of the circuit, the detection of antiphase mode becomes the main working state. In addition, as if to take $\Delta\omega_I =$

Table 5.1 Actual parameters of GY-027 gyro detection mode

Parameter name	Parameter value
Driving mode resonant frequency ω_d	$3488.9 \times 2\pi$ rad/s
Sensing in-phase mode resonance frequency ω_{y1}	$3360.1 \times 2\pi$ rad/s
Sensing anti-phase mode resonant frequency ω_{y2}	$3464.1 \times 2\pi$ rad/s
Sensing in-phase mode quality factor Q_{y1}	1051
Sensing anti-phase mode quality factor Q_{y2}	1224

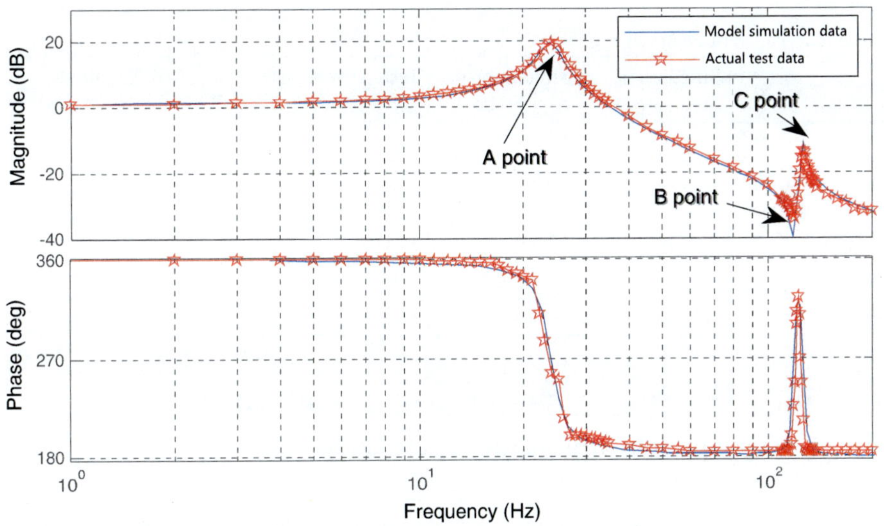

Fig. 5.5 Test and simulation diagram of dual-mass linear vibration silicon-based MEMS gyroscope sensing mode transfer function

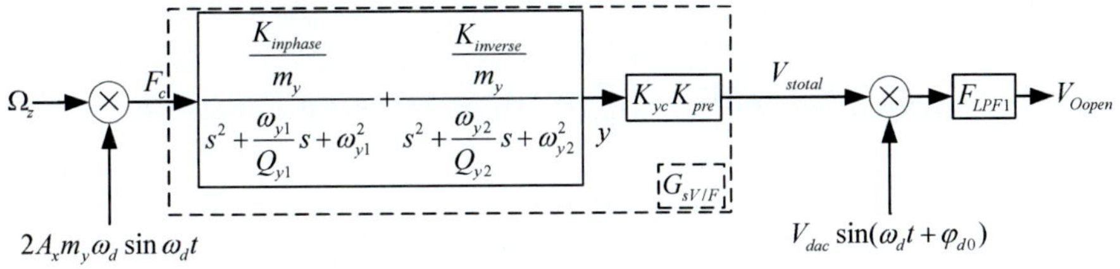

Fig. 5.6 Block diagram of dual-mass gyro sensing mode open loop system

$|\omega_{y1}\text{-}\omega_d|$, $\Delta\omega_2 = |\omega_{y2}\text{-}\omega_d|$, the bandwidth of the detection mode should be determined by the frequency difference of the smaller (as if $\Delta\omega_1 < \Delta\omega_2$; The resonant peak of $\Delta\omega 2$ and $\Delta\omega 1$ may be less than 3 dB, and the bandwidth should be determined by B valley or $\Delta\omega_2$ peak). Table 5.2 lists the data of each monitoring point in Fig. 5.5. It can be seen from the table that, at each key monitoring point, the simulation results of the detection system model are basically consistent with the measured results, which further proves that the dual-quality detection mode model is formed by the superposition of in-phase and anti-phase modes. At the same time, this model also lays a solid foundation for the design of open-loop and closed-loop detection loops in the following chapters.

Table 5.2 The parameters in Fig. 5.5

	Simulation Frequency	Simulation Amplitude	Tested Frequency	Tested Amplitude
DC Point	-	1.15 dB	-	1.07 dB
Bandwidth Point(+3dB)	12.9 Hz	4.15 dB	13 Hz	4.11 dB
Point A	24.8 Hz	19.5 dB	24 Hz	19.7 dB
Point B	120 Hz	−39.3 dB	120 Hz	−34.4 dB
Point C	128 Hz	−11.4 dB	128 Hz	−13.9 dB

5.4 Dual-Mass Linear Vibration Silicon-Based MEMS Gyroscope Sensing Mode Open Loop

The sensing open loop is used to process the output signal of the sensing mode interface circuit, and the Coriolis signal is usually extracted by the method of phase sensitive demodulation. In addition, the sensing open loop is also a part of the detection closed-loop loop in the subsequent research.

5.4.1 Sensing Mode Open Loop System Working Principle

In the case of ignoring various external noises, the block diagram of the sensing open loop system after the superposition of sensing mode of the dual-mass linear vibration silicon-based MEMS gyroscope is shown as follows (Fig. 5.6).

In the figure, $V_{dac} = 1.8$ V is the demodulation reference amplitude; φ_{d0} is demodulation reference phase; V_{stotal} is the signal before demodulation. The following equations can be extracted from the figure above:

$$F_c(t) = 2\Omega_z(t) m_y A_x \omega_d \sin \omega_d t \tag{5.4}$$

$$G_{sV/F} = \left(\frac{K_{inphase}}{s^2 + \frac{\omega_{y1}}{Q_{y1}} s + \omega_{y1}^2} + \frac{K_{inverse}}{s^2 + \frac{\omega_{y2}}{Q_{y2}} s + \omega_{y2}^2} \right) K_{yc} K_{pre} \tag{5.5}$$

$$V_{stotal}(s) = F_c(s) G_{sV/F} \tag{5.6}$$

$$V_{Oopen} = V_{dac} \sin(\omega_d t + \varphi_{d0}) V_s F_{LPF1} \tag{5.7}$$

Apply Laplace transform to Eq. (5.4), and combine Eqs. (5.5) and (5.6) as follows:

$V_{stotal}(s)$

$$= A_x\omega_d K_{yc} K_{pre}(\Omega_z(s - j\omega_d) + \Omega_z(s + j\omega_d))\left(\frac{K_{inphase}}{s^2 + \frac{\omega_{y1}}{Q_{y1}}s + \omega_{y1}^2} + \frac{K_{inverse}}{s^2 + \frac{\omega_{y2}}{Q_{y2}}s + \omega_{y2}^2}\right) \tag{5.8}$$

Substitute Eq. (5.8) into Eq. (5.7):

$$\begin{aligned} V_{Oopen}(s) &= \frac{1}{2}A_x\omega_d V_{dac} K_{yc} K_{pre} F_{LPF1}(s) \\ &\left\{\left(\frac{[\Omega_z(s) + \Omega_z(s + 2j\omega_d)]e^{-j\varphi_{d0}}K_{inphase}}{(s + j\omega_d)^2 + \frac{\omega_{y1}}{Q_{y1}}(s + j\omega_d) + \omega_{y1}^2} + \frac{[\Omega_z(s) + \Omega_z(s + 2j\omega_d)]e^{-j\varphi_{d0}}K_{inverse}}{(s + j\omega_d)^2 + \frac{\omega_{y2}}{Q_{y2}}(s + j\omega_d) + \omega_{y2}^2}\right) + \right. \\ &\left.\left(\frac{[\Omega_z(s - 2j\omega_d) + \Omega_z(s)]e^{j\varphi_{d0}}K_{inphase}}{(s - j\omega_d)^2 + \frac{\omega_{y1}}{Q_{y1}}(s - j\omega_d) + \omega_{y1}^2} + \frac{[\Omega_z(s - 2j\omega_d) + \Omega_z(s)]e^{j\varphi_{d0}}K_{inverse}}{(s - j\omega_d)^2 + \frac{\omega_{y2}}{Q_{y2}}(s - j\omega_d) + \omega_{y2}^2}\right)\right\} \end{aligned} \tag{5.9}$$

The low-pass filter F_{LPF1} filters out the high-frequency components and ignores the demodulation phase angle error, namely $\varphi_{d0} = 0$, and the formula above is:

$$\left|\frac{V_{Oopen}(s)}{\Omega_z(s)}\right| = \left|\frac{1}{2}A_x\omega_d V_{dac} K_{yc} K_{pre} F_{LPF1}(s) G_{equal}(s)\right| \tag{5.10}$$

Among them:

$$G_{equal}(s) = \left\{\frac{K_{inphase}(s^2 + \frac{\omega_{y1}}{Q_{y1}}s + \omega_{y1}^2 - \omega_d^2)}{(s^2 + \frac{\omega_{y1}}{Q_{y1}}s + \omega_{y1}^2 - \omega_d^2)^2 + (2s\omega_d + \frac{\omega_{y1}}{Q_{y1}}\omega_d)^2} + \frac{K_{inverse}(s^2 + \frac{\omega_{y2}}{Q_{y2}}s + \omega_{y2}^2 - \omega_d^2)}{(s^2 + \frac{\omega_{y2}}{Q_{y2}}s + \omega_{y2}^2 - \omega_d^2)^2 + (2s\omega_d + \frac{\omega_{y2}}{Q_{y2}}\omega_d)^2}\right\}$$

In Eq. (5.10), in addition to the poles of $F_{LPF1}(s)$, the system also has the following eight poles:

$$p_{1,2} = -\frac{\omega_{y1}}{2Q_{y1}} + (\omega_d \pm \frac{\omega_{y1}}{2}\sqrt{4 - \frac{1}{Q_{y1}^2}})j;\ p_{3,4} = -\frac{\omega_{y1}}{2Q_{y1}} - (\omega_d \pm \frac{\omega_{y1}}{2}\sqrt{4 - \frac{1}{Q_{y1}^2}})j$$

.

$$p_{5,6} = -\frac{\omega_{y2}}{2Q_{y2}} + (\omega_d \pm \frac{\omega_{y2}}{2}\sqrt{4 - \frac{1}{Q_{y2}^2}})j;\ p_{7,8} = -\frac{\omega_{y2}}{2Q_{y2}} - (\omega_d \pm \frac{\omega_{y2}}{2}\sqrt{4 - \frac{1}{Q_{y2}^2}})j$$

.

These poles are in the plane of negative real axis, so the system is stable. Because the structure exists in the vacuum package, $Q_{y1} >> 1$, $Q_{y2} >> 1$, so the poles above can be simplified as:

$$p_{1,2,3,4} = -\frac{\omega_{y1}}{2Q_{y1}} \pm (\omega_d \pm \omega_{y1})j;\ p_{5,6,7,8} = -\frac{\omega_{y2}}{2Q_{y2}} \pm (\omega_d \pm \omega_{y2})j$$

.

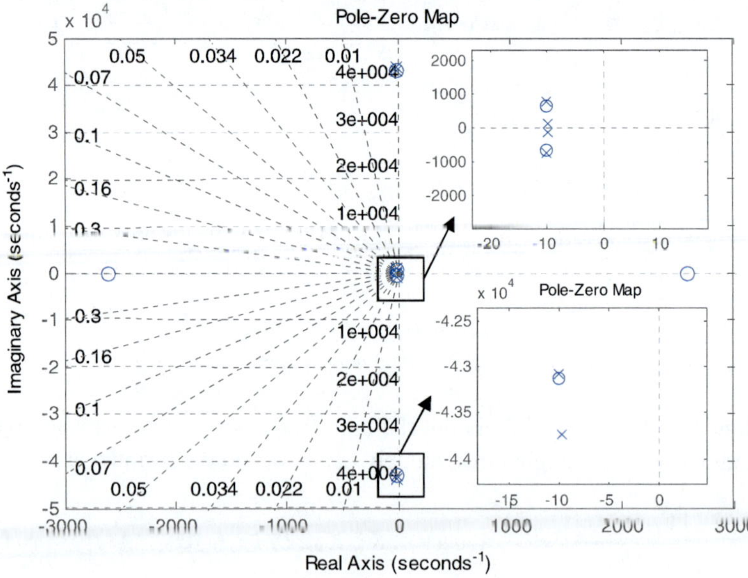

Fig. 5.7 Distribution of zero and pole of sensing open loop

The distribution of zero poles and poles in the open loop state of GY-027 gyro was obtained by substituting all parameters into the model, as shown in Fig. 5.7. It can be seen from the figure that the distribution of poles of the system is the same as the conclusion of the above theoretical analysis, and the pole closest to the origin determines the bandwidth of GY-027.

5.4.2 Sensing Mode Open Loop System Bandwidth

After the parameters are substituted into Formula (5.10), the simulation curve of the sensing open loop system with high frequency band can be obtained as shown in Fig. 5.8. The difference between the low frequency band in the figure and Fig. 5.5 is due to the different input signals of the Fig. 5.8. The input signal in Fig. 5.8 is angular rate Ω_z, while the input signal in Fig. 5.5 is the force of the mass block. According to Eq. (5.10), the steady-state scaling factor is:

$$\left|\frac{V_{Oopen}(0)}{\Omega_z(0)}\right| = \left|\frac{1}{2}A_x\omega_d V_{dac}K_{yc}K_{pre}F_{LPF1}(0)\left\{\frac{K_{inphase}(\omega_{y1}^2-\omega_d^2)}{(\omega_{y1}^2-\omega_d^2)^2+(\frac{\omega_{y1}}{Q_{y1}}\omega_d)^2}+\frac{K_{inverse}(\omega_{y2}^2-\omega_d^2)}{(\omega_{y2}^2-\omega_d^2)^2+(\frac{\omega_{y2}}{Q_{y2}}\omega_d)^2}\right\}\right| \quad (5.11)$$

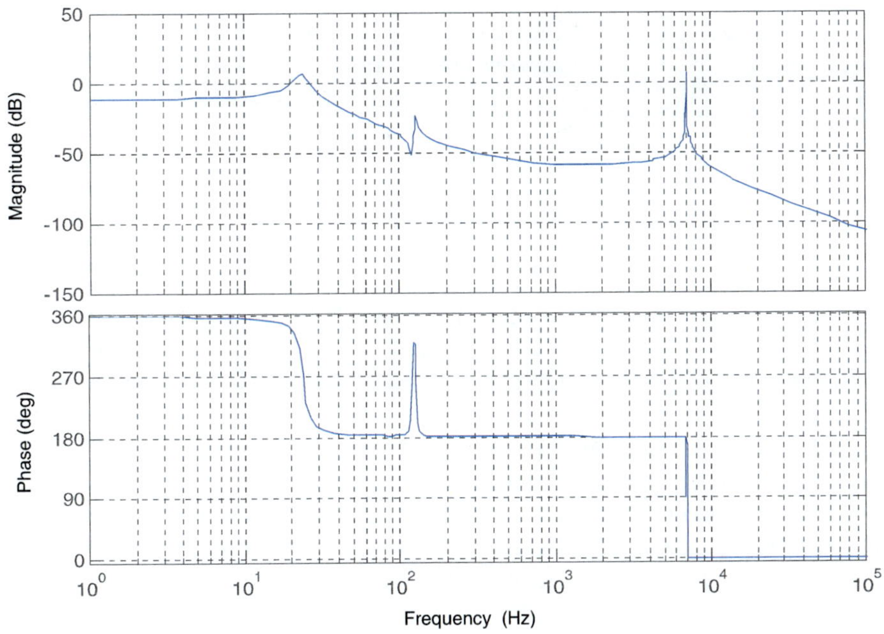

Fig. 5.8 Potter figure of sensing open loop system

Because $(\omega_{y1,2}^2 - \omega_d^2) \gg \frac{\omega_{y1,2}}{Q_{y1,2}}\omega_d$, and the previous section proves that there is $K_{inverse} >> K_{inphase}$ in the actual sensing model which mainly sensing anti-phase mode, then Eq. (5.11) can be simplified as:

$$\left|\frac{V_{Oopen}(0)}{\Omega_z(0)}\right| = \left|\frac{A_x V_{dac} K_{yc} K_{pre} K_{inverse} F_{LPF1}(0)}{2\Delta\omega_2}\right| \tag{5.12}$$

In addition, the output amplitude of its sensing open loop can be obtained, and the relevant parameters of GY-027 gyro can be substituted and simplified. Then the scale factor of the gyro can be expressed as:

$$\left|\frac{V_{Oopen}(s)}{\Omega_z(s)}\right| = \frac{A_x V_{dac} K_{yc} K_{pre} K_{inverse} F_{LPF1}(s)(s^2 + \frac{\omega_{y2}}{Q_{y2}}s + \omega_{y2}^2 - \omega_d^2)}{(s^2 + \frac{\omega_{y2}}{Q_{y2}}s + \omega_{y2}^2 - \omega_d^2)^2 + (2s\omega_d + \frac{\omega_{y2}}{Q_{y2}}\omega_d)^2} \tag{5.13}$$

In gyro bandwidth, a certain angular frequency $\omega_{ope<<}\omega_{y2}$ corresponding scale factors are:

$$\left|\frac{V_{Oopen}(\omega_{open})}{\Omega_z(\omega_{open})}\right| = \frac{A_x V_{dac} K_{yc} K_{pre} K_{inverse} F_{LPF1}(\omega_{open})}{2\Delta\omega_2\left(1 - \frac{\omega_{open}^2}{\Delta\omega_2^2}\right)} \tag{5.14}$$

For the gyro of $\omega_{y1} < \omega_{y2} < \omega_d$ (the bandwidth is determined by point A in Fig. 5.5), whose bandwidth restriction conditions are as follows:

$$\left| \frac{V_{Oopen}(\omega_b)}{\Omega_z(\omega_b)} \right| = \sqrt{2} \left| \frac{V_{Oopen}(0)}{\Omega_z(0)} \right| \tag{5.15}$$

It can be solved according to Eqs. (5.12) and (5.14), so that (F_{LPF1} is constant in gyro bandwidth):

$$\omega_b = 0.54\Delta\omega_2 \tag{5.16}$$

For GY-027, $\Delta\omega_2 = 24.8$ Hz, then the corresponding bandwidth can be calculated as 13.4 Hz through the above formula, which is basically consistent with the data in Table 5.2, proving the accuracy of the above analysis. Since the mode frequency difference of the silicon-based MEMS gyroscope is fixed after the structural machining is completed, and it can hardly be greatly changed in the absence of the teeth that produce the mode negative stiffness, the bandwidth of the gyro in the open loop state is almost fixed (excluding the change of frequency difference caused by temperature and other factors).Since the mechanical sensitivity and scale factor of the gyro are inversely proportional to $\Delta\omega$, but the bandwidth is proportional to it, the high sensitivity needs to be sacrificed in the sensing open loop scheme to ensure the working bandwidth of the gyro. In order to satisfy these two requirements, it is necessary to study the sensing closed loop system and find the method of bandwidth expansion from the perspective of control system.

5.4.3 *Sensing Mode Open Loop System Simulation*

Combined with the previous chapters, the sensing open loop is established in the Simulink simulation environment as shown in Fig. 5.9, in which the orthogonal coupling stiffness correction method is used in the orthogonal correction, and the control loop is integrated in the sub-module. In the figure, after the system is powered on and stabilized, step input angle rate $\Omega_z = 100°/\text{s}$, and the system output V_{Oopen} and detection displacement signal are observed, as shown in Fig. 5.10. Within 0.5 s after the angular rate input, the gyroscope has formed a stable Coriolis signal output, and the maximum displacement of the corresponding sensing mode is about 0.04 um.

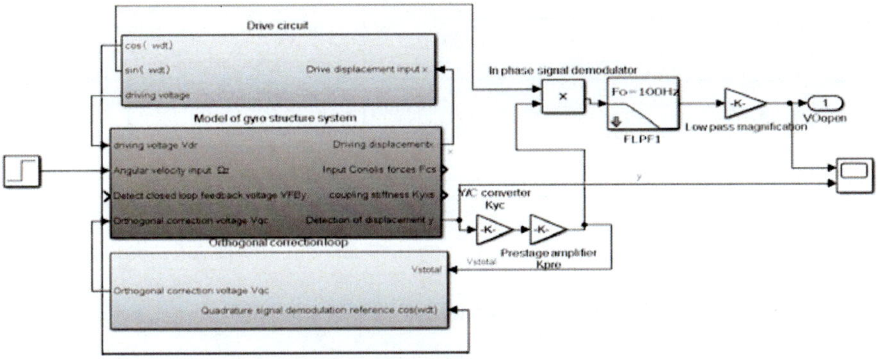

Fig. 5.9 Simulation diagram of sensing open loop system

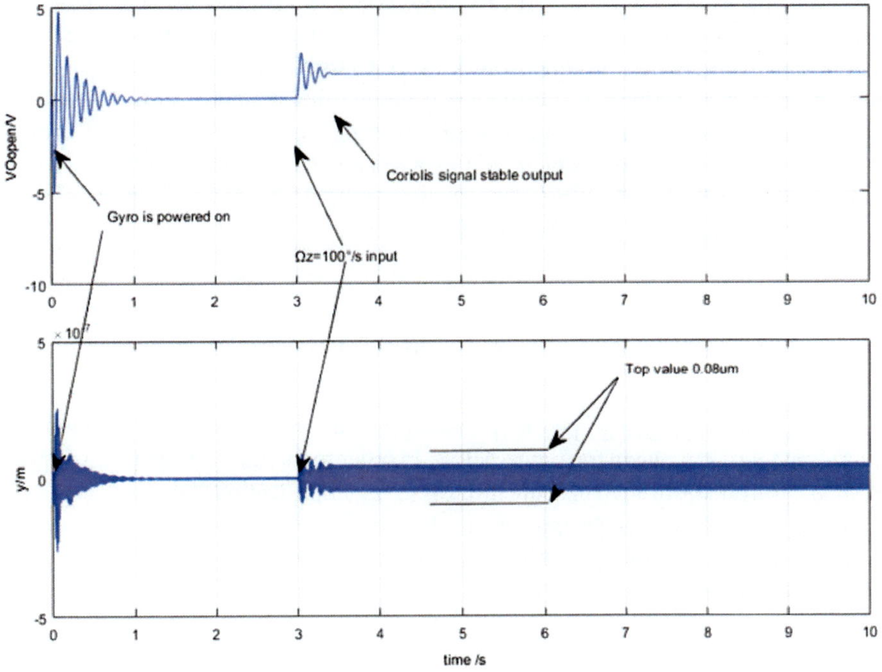

Fig. 5.10 Simulation diagram of output curve of sensing open loop

5.5 Dual-Mass Linear Vibration Silicon-Based MEMS Gyroscope Sensing Mode Closed Loop

A more widely used silicon-based MEMS gyroscope must have both good static and dynamic performance. The static characteristics are mainly reflected in the scale factor, noise level and zero bias stability of gyro. The dynamic characteristics of gyro

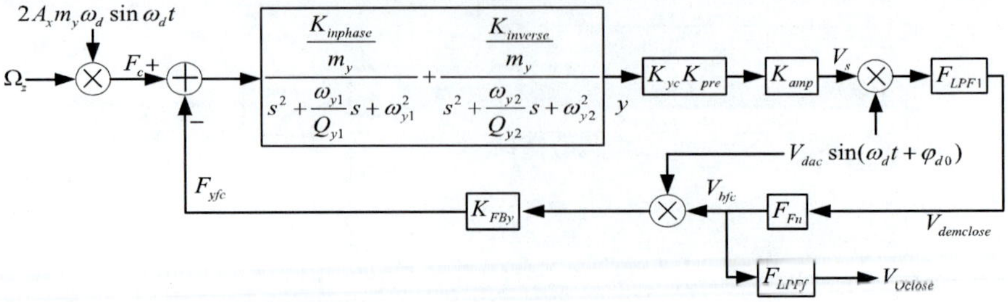

Fig. 5.11 Block diagram of dual-mass gyro detection closed-loop system

are mainly investigated in the bandwidth, anti-vibration, impact and other aspects of the ability, as well as the nonlinearity of the scale factor. When the gyro sensing loop is in closed loop state, the sensing force feedback comb will provide the corresponding static force to force the sensing mass to be in a static state in the sensing direction. At the same time, when the mass block is in a large impact and overload state, the sensing force feedback static force can protect the structure from damage. In addition, the sensing open loop adopts the working mode of converting Coriolis force into comb tooth displacement and then sensing capacitance, while the closed loop directly senses Coriolis force, thus avoiding the influence of nonlinear factors such as comb tooth sensing in the conversion process.

5.5.1 Sensing Mode Closed Loop System Working Principle

Combined with the sensing open loop in the previous section, the working mode of the closed loop is shown in the figure below. In order to facilitate the gain adjustment of the sensing loop in the subsequent design, an amplification link K_{amp} is added after the output of the pre-sensing stage. Where, F_{Fn} is the feedback force controller; F_{LPFf} is a closed loop low-pass filter; V_{bfc} is the feedback control voltage. It is not difficult to obtain the equations (Fig. 5.11):

$$y(s)=\left(F_c(s)-F_{yfc}(s)\right)\left(\frac{K_{inphase}}{s^2+\frac{\omega_{y1}}{Q_{y1}}s+\omega_{y1}^2}+\frac{K_{inverse}}{s^2+\frac{\omega_{y2}}{Q_{y2}}s+\omega_{y2}^2}\right) \tag{5.17}$$

$$V_s(s)=y(s)K_{yc}K_{pre}K_{amp} \tag{5.18}$$

$$V_{bfc}(t)=V_{demclose}(t)F_{Fn} \tag{5.19}$$

$$F_{yfc}(t)=K_{FBy}V_{bfc}(t)V_{dac}\sin(\omega_d t+\varphi_{d0}) \tag{5.20}$$

$$V_{Oclose} = V_{bfc} F_{LPFf} \tag{5.21}$$

By referring to the conversion mode in the above section and ignoring the influence of demodulation phase angle ($\varphi_{d0} = 0$), the time-domain signal in the above formula is converted to the frequency-domain signal. After the action of demodulator and low-pass filter F_{LPF1}, the following results can be obtained:

$$V_{demclose}(s) = \frac{V_{dac}}{2} K_{yc} K_{pre} K_{amp} F_{LPF1}(s)\left(A_x \omega_d \Omega_z(s) - \frac{V_{dac} K_{FBy} V_{bfc}(s)}{2m_y}\right) G_{equal}(s) \tag{5.22}$$

The scaling factor of the closed loop system can be obtained by subsuming the Laplace transform of Eq. (5.19) into Eq. (5.22):

$$\left|\frac{V_{Oclose}(s)}{\Omega_z(s)}\right| = \left|\frac{2V_{dac} K_{yc} K_{pre} K_{amp} F_{LPF1}(s) F_{Fn}(s) F_{LPFf}(s) A_x \omega_d G_{equal}(s)}{4 + \frac{V_{dac}^2 K_{FBy}}{m_y} K_{yc} K_{pre} K_{amp} F_{LPF1}(s) F_{Fn}(s) G_{equal}(s)}\right| \tag{5.23}$$

The above formula can be described by the system block diagram below.

5.5.2 Sensing Mode Closed Loop System Bandwidth

According to the F_{Fn} value in the denominator of Eq. (5.23), the closed loop system scale factor has the following two states:

1. If F_{Fn} is very small, there is $\frac{V_{dac}^2 K_{FBy}}{4m_y} K_{yc} K_{pre} K_{amp} F_{LPF1}(s) F_{Fn}(s) G_{equal}(s) \ll 1$, then:

$$\left|\frac{V_{Oclose}(s)}{\Omega_z(s)}\right| = \left|\frac{1}{2} V_{dac} K_{yc} K_{pre} K_{amp} F_{LPF1}(s) F_{Fn}(s) F_{LPFf}(s) A_x \omega_d G_{equal}(s)\right| \tag{5.24}$$

After ignoring the action of in-phase components sensed in $G_{equal}(s)$, parameters can be substituted to further simplify the above equation:

$$\left|\frac{V_{Oclose}(s)}{\Omega_z(s)}\right| = \left|\frac{A_x \omega_d K_{inverse} V_{dac} K_{yc} K_{pre} K_{amp} F_{LPF1}(s) F_{Fn}(s) F_{LPFf}(s)(s^2 + \frac{\omega_{y2}}{Q_{y2}} s + \omega_{y2}^2 - \omega_d^2)}{(s^2 + \frac{\omega_{y2}}{Q_{y2}} s + \omega_{y2}^2 - \omega_d^2)^2 + (2s\omega_d + \frac{\omega_{y2}}{Q_{y2}} \omega_d)^2}\right| \tag{5.25}$$

Suppose ω_{close} is the angular frequency at some point in the bandwidth, have $\omega_{close} << \omega_d$, then the scale factor at this frequency point can be further simplified as:

$$\left|\frac{V_{Oclose}(\omega_{close})}{\Omega_z(\omega_{close})}\right| = \left|\frac{A_x K_{inverse} V_{dac} K_{yc} K_{pre} K_{amp} F_{LPF1}(\omega_{close}) F_{Fn}(\omega_{close}) F_{LPFf}(\omega_{close})}{2\Delta\omega_2\left(1 - \frac{\omega_{close}^2}{\Delta\omega_2^2}\right)}\right| \tag{5.26}$$

It can be seen that when the force feedback controller F_{Fn} is very small, the scale factor of the closed loop system is still mainly determined by the frequency difference of the modes, and its value is similar to that of the open loop (Eq. 5.12). Its bandwidth is determined by the following formula:

$$\omega_{bclose} = \Delta\omega_2\sqrt{1 - \frac{\sqrt{2}F_{Fn}(\omega_{bclose})}{2F_{Fn}(0)}} \tag{5.27}$$

So, to get omega $b_{close} > \omega_b$, then $F_{Fn}(\omega_{bclose}) < F_{Fn}\ (0)$.However, since the above method is based on the small F_{Fn}, the feedback force generated by it is not enough to limit the motion amplitude of the sensing quality, and its ability to expand bandwidth is limited. In essence, this method is an optimized version of the sensing open loop method and cannot effectively play the advantages of the sensing closed loop.

2. When F_{Fn} is large, there is $\frac{V_{dac}^2 K_{FBy}}{4m_y} K_{yc} K_{pre} K_{amp} F_{LPF1}(s) F_{Fn}(s) G_{equal}(s) \gg 1$, according to Eq. (5.23), there is:

$$\left|\frac{V_{Oclose}(s)}{\Omega_z(s)}\right| = \left|\frac{2m_y F_{LPFf}(s) A_x \omega_d}{K_{FBy} V_{dac}}\right| \tag{5.28}$$

It can be seen from the above equation that the scale factor of the closed loop is proportional to the vibration amplitude and natural frequency of the driving mode, and the bandwidth is only determined by the low-pass filter of the output stage. In addition, too large feedback controller coefficient will make the noise system unstable. By substituting the actual parameters of gyro into Fig. 5.12, the distribution diagram of zero and pole of the whole closed loop system can be drawn according to different F_{Fn} values, so as to calculate the value range of F_{Fn} values. Set F_{Fn} values of 0.1, 1, 10, 50, 100, 200, 300, 400, 500, 1000, 2000, 5000 and 10,000 respectively to obtain the corresponding zero-pole distribution diagram of the closed loop system, as shown in Fig. 5.13 (since the change of F_{Fn} value has almost no influence on the zero-pole distribution at high frequency, therefore, this figure focuses on the distribution near the origin.

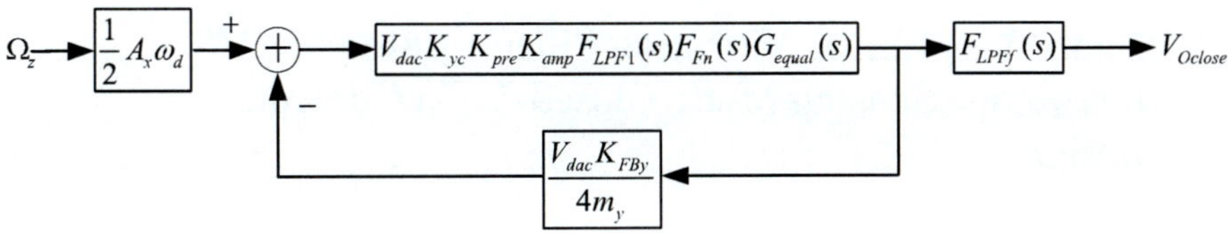

Fig. 5.12 Block diagram of sensing closed loop system

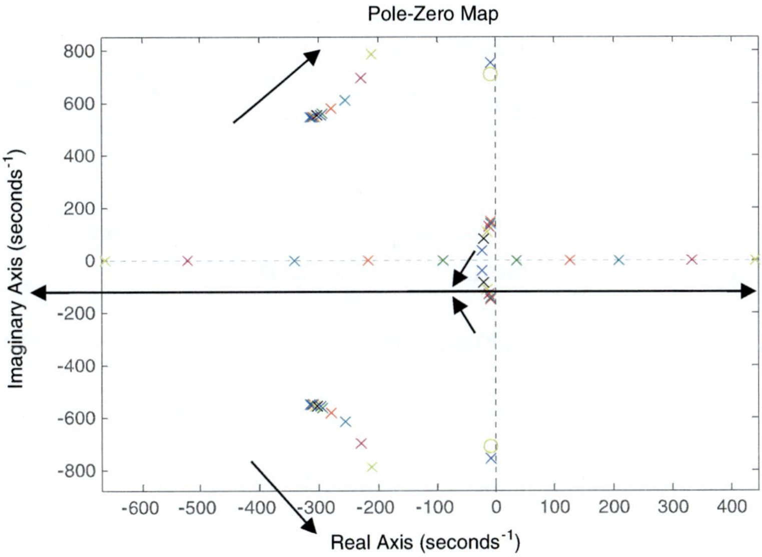

Fig. 5.13 Influence of F_{Fn} value change on zero-pole distribution of closed loop system

The black arrow line in the figure indicates that with the increase of F_{Fn} value, the dominant pole of the system (in the S-plane, the closed loop pole closest to the virtual axis and without closed loop zero distribution nearby, has the greatest influence on the performance of the system. At the same time, the influence of closed loop zero poles with real parts more than six times larger than the dominant poles on the system is negligible [147]) distribution trend of it. When the F_{Fn} value is close to 500, the poles on the positive half axis of the real axis appear, which indicates that the system is in an unstable state, and the poles on the positive half axis of the real axis continue to increase with the increase of F_{Fn}, so the simple increment controller cannot meet the needs of closed loop control stability, and it is necessary to design the force feedback controller in detail.

5.6 Dual-Mass Linear Vibration Silicon-Based MEMS Gyroscope Sensing Mode Closed-Loop Controller Design

In order to achieve better signal-to-noise ratio and mechanical sensitivity for the structure of dual-mass linear vibration silicon-based MEMS gyroscope, narrow-frequency difference gyro is selected in this chapter. As shown in Eq. (5.16), mechanical band width limits the overall bandwidth of gyro. Therefore, the closed loop controller should have the ability of bandwidth expansion. In engineering, the working bandwidth of gyro is usually required to be greater than 50 Hz. This part of the book will be one of the research focuses.

5.6.1 Sensing Mode Closed Loop Controller Design Based on Pole-Zero Compensation Method

5.6.1.1 Working Principle of Dipole Compensation Controller

The principle of automatic control states that if the distance between zero and pole is one order of magnitude smaller than their own modulus, they constitute a dipole, and the influence of the dipole far from the origin can be ignored [147]. Figure 5.14 shows the dipole.

In the figure p_{ri} is the mode of the pole. According to the analysis in the preceding sections, the silicon-based MEMS gyroscope closed loop bandwidth is completely determined by the dominant pole of the closed loop system (the pole nearest to the origin, generated by sensing anti-phase mode) under the condition that the phase factor remains unchanged. Therefore, the system bandwidth can be extended by configuring the corresponding system zero and the dominant pole to form a dipole. The leading poles of gyro sensing mode are:

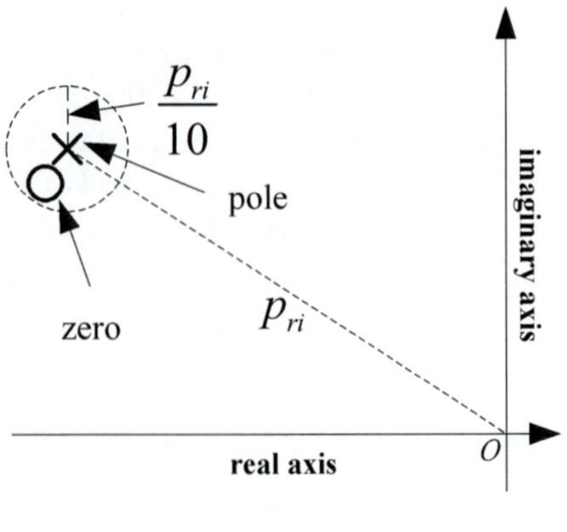

Fig. 5.14 Schematic diagram of dipole

$$p_{5,6} = -\frac{\omega_{y2}}{2Q_{y2}} \pm \Delta\omega_2 j \tag{5.29}$$

In order to minimize the influence of the two conjugate dominant poles in the above formula and increase the flatness in gyro bandwidth, the zero point of controller FFn should be selected to be equal to the poles in the above formula, namely, $z5,6 = p5,6$.

5.6.1.2 Dipole Compensation Controller Design

After determining the zero point of the controller, it is necessary to design the controller accurately. The controller should be composed of simple analog circuit modules, and in order to enhance the stability of the system, the denominator of the controller should also be realized by a second-order system. In addition, the pole ω_{Fn} of the controller itself should be located outside the gyro bandwidth frequency, otherwise the purpose of bandwidth expansion cannot be achieved. Based on the above analysis, it can be concluded that:

$$F_{Fn}(s) = \frac{\left(s + \frac{\omega_{y2}}{2Q_{y2}} + \Delta\omega_2 j\right)\left(s + \frac{\omega_{y2}}{2Q_{y2}} - \Delta\omega_2 j\right)}{A_{Fn}(s + \omega_{Fn})^2} \tag{5.30}$$

where, A_{Fn} is the coefficient of the denominator of the controller, which can be further expanded as follows:

$$F_{Fn}(s) = \frac{s^2 + \frac{\omega_{y2}}{Q_{y2}}s + \Delta\omega_2^2 + \frac{\omega_{y2}^2}{4Q_{y2}^2}}{A_{Fn}(s + \omega_{Fn})^2} \tag{5.31}$$

In general, the driving and sensing anti-phase mode of the gyro do not match, and the gyro structure is in a high vacuum package, so:

$$\Delta\omega_2^2 \gg \frac{\omega_{y2}^2}{4Q_{y2}^2}. \tag{5.32}$$

Substitute the above equation into Eq. (5.31), then:

$$F_{Fn}(s) = \frac{1}{A_{Fn}} \frac{s^2 + \frac{\omega_{y2}}{Q_{y2}}s + \Delta\omega_2^2}{s^2 + 2\omega_{Fn}s + \omega_{Fn}^2} \tag{5.33}$$

Based on the analysis in the previous chapters, the parameters in the above formula should meet the following conditions:

1. Pole distribution should be greater than the required bandwidth of gyro (50 Hz):

$$\omega_{Fn} \geq 50 * 2 * \pi \tag{5.34}$$

2. F_{Fn} gain should be kept within 400:

$$\frac{\Delta\omega_2^2}{A_{Fn}\omega_{Fn}^2} \leq 400 \tag{5.35}$$

3. Since the poles of conjugate complex numbers will produce the peaks of amplitude frequency characteristics in the Bode diagram, the poles of the controller should be the real roots to ensure that the amplitude-frequency curve of the system rapidly attenuates with a slope of −40 dB, which is conducive to reducing the influence of high-frequency noise.

Based on the above conditions, this section adopts operational amplifier to build closed loop controller, as shown in Fig. 5.15 [148]. The transfer function of the circuit in the figure can be expressed as follows:

$$F_{Fn}(s) = \frac{V_{fo}}{V_{fi}} = \frac{R_{f12}R_{f7}}{R_{f11}R_{f8}} \frac{s^2 + \frac{R_{f8}R_{f4}}{R_{f5}R_{f9}R_{f10}C_{f2}}s + \frac{R_{f8}}{R_{f1}R_{f2}R_{f9}C_{f1}C_{f2}}}{s^2 + \frac{R_{f4}R_{f7}}{R_{f6}R_{f9}R_{f10}C_{f2}}s + \frac{R_{f7}}{R_{f2}R_{f3}R_{f9}C_{f1}C_{f2}}} \tag{5.36}$$

where, the equivalent resistance $R_{f1} = \frac{R_{f1a}R_{f1c}}{R_{f1b}+R_{f1a}}$, let $R_{f1b} >> R_{fab}$, then $R_{f1} = \frac{R_{f1a}R_{f1c}}{R_{f1b}}$.

Combining Eqs. (5.36) and (5.33), we have:

$$\frac{\omega_{y2}}{Q_{y2}} = \frac{R_{f8}R_{f4}}{R_{f5}R_{f9}R_{f10}C_{f2}},\ \Delta\omega_2^2 = \frac{R_{f8}}{R_{f1}R_{f2}R_{f9}C_{f1}C_{f2}},\ 2\omega_{Fn} = \frac{R_{f4}R_{f7}}{R_{f6}R_{f9}R_{f10}C_{f2}}.$$

Substitute the data in Table 5.1 into the above equation to obtain the resistance and capacity parameters in Fig. 5.15. When in DC state:

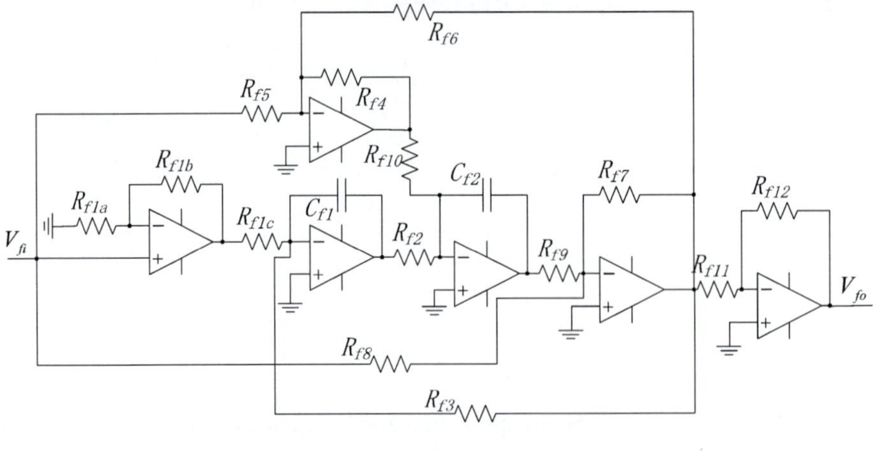

Fig. 5.15 Circuit diagram of sensing closed loop controller

Table 5.3 F_{Fn} Actual resistance and capacity parameters of the controller

Parameters of the label	Parameter values	Parameters of the label	Parameter values
R_{f1}	1400 Ω	R_{f8}	100 Ω
R_{f2}	15000 Ω	R_{f9}	100 Ω
R_{f3}	100 Ω	R_{f10}	1000 Ω
R_{f4}	20000 Ω	R_{f11}	1000 Ω
R_{f5}	100000 Ω	R_{f12}	100000 Ω
R_{f6}	1600 Ω	C_{f1}	0.2 uF
R_{f7}	500000 Ω	C_{f2}	10 uF

$$F_{Fn} = \frac{R_{f3}R_{f12}}{R_{f1}R_{f11}} \tag{5.37}$$

Because $\omega_{Fn} > \Delta\omega_2$, so $R_{f3}/\ R_{F1} < 1$. Gain characteristics required by the closed loop controller can be achieved by configuring the corresponding Rf12/Rf11 ratio. Table 5.3 lists the resistance and capacity parameters (5.37).

By substituting all parameters in Table 5.3 into Eq. (5.36), the simulated Bode diagram of $F_{Fn}(s)$ can be obtained, as shown in the solid line below. In order to grasp the circuit characteristics in Fig. 5.15 more directly and accurately, Multisim software is used to simulate it, and the simulation results are shown in Fig. 5.16. It can be seen that the simulation results of controller transmission and circuit basically coincide, so the design circuit can accurately reflect the transfer function.

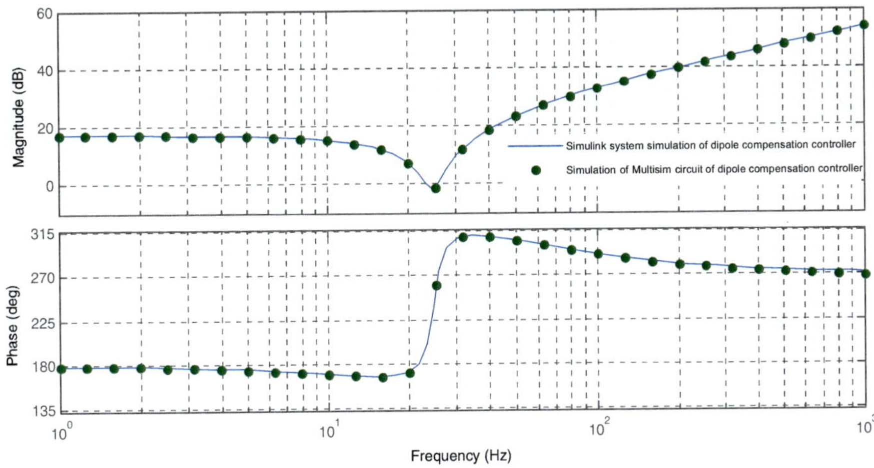

Fig. 5.16 Circuit simulation of dipole compensation controller

5.6.1.3 Dipole Compensation Control System Time Domain Simulation

According to the designed controller parameters, a feedback controller module is added on the basis of Fig. 5.9, as shown in Fig. 5.17. Also, after 1 s of power-on, add angular rate $\omega_Z = 100°/s$ to detect gyro output and sensing displacement, and the results are shown in the figure below. It can be seen from the figure that the gyro can reach a stable state after it is powered on for about 0.5 s, and the gyro output is stable after the angular velocity is input for 0.5 s. In addition, due to the action of the feedback torturer, the sensing mode displacement decreases by nearly two orders of magnitude compared with the open loop state. The gyroscope output signal size can be controlled by the output stage amplifier (Fig. 5.18).

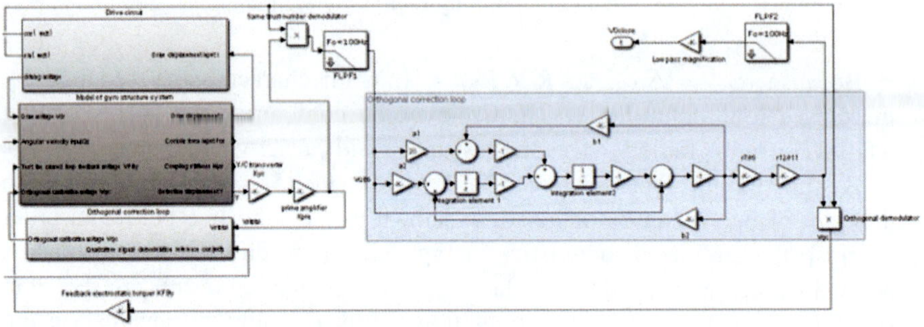

Fig. 5.17 Dipole compensation control system model

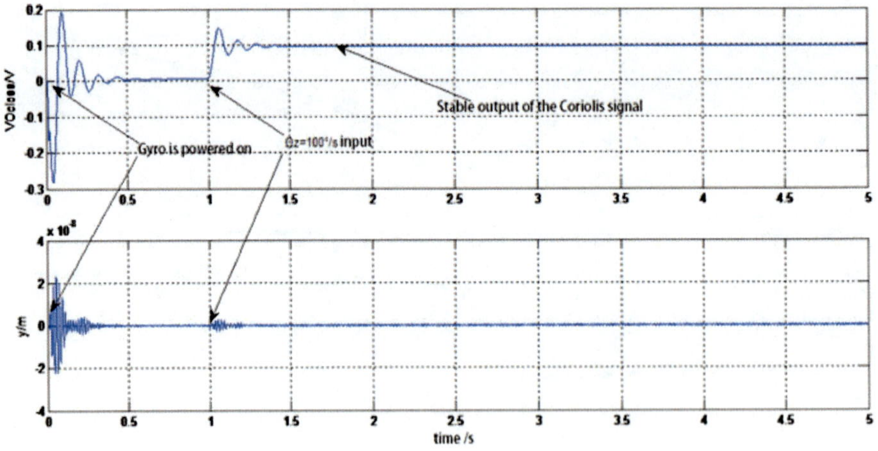

Fig. 5.18 Time domain simulation of dipole compensation control system

5.6.1.4 Frequency Domain Simulation of Dipole Compensation Control System

On the basis of the above analysis, the overall simulation diagram of the sensing closed loop system can be obtained, as shown below.

After plugging in all parameters, simulation was carried out on Fig. 5.19, focusing on the observation of its closed loop zero-pole distribution and amplitude-frequency and phase-frequency curves, as shown in Figs. 5.20 and 5.21. Among them, the closed loop zero-pole distribution diagram mainly shows the position of the dominant pole and the zero point in the low frequency band. All poles of the closed loop system are located in the negative half axis of the real axis, which proves the stability of the system. The dominant pole also coincides with the zero point of the controller, forming a dipole. The system bode diagram shows that the bandwidth of the closed loop system is 63.5 Hz, which meets the design requirements. At the same time, the resonant peak caused by the gyro frequency difference is basically eliminated due to the good match between the controller zero and the dual-mass linear vibration silicon-based MEMS gyroscope mode poles, and the in-band flatness is ideal. In addition, in Fig. 5.20, a pair of conjugate complex poles generated by sensing in-phase mode and driving anti-phase mode frequency difference ($\Delta\omega_1$) does not form a dipole with its nearby zero point, which directly results in "valleys" and "peaks" outside the bandwidth in Fig. 5.21 (points B and C in Fig. 5.5). As the current bandwidth has met the design requirements of 50 Hz, there is no need to deal with it. However, if the gyro needs a larger bandwidth (such as more than 100 Hz), the influence of this factor must be paid attention to in the design stage of the dual-mass linear vibration silicon-based MEMS gyroscope structure.

The closed loop Quist curve of the system is shown in Fig. 5.22. The curve in the figure does not contain the point (0, −1j). The system can be judged to be in a stable state according to relevant criteria.

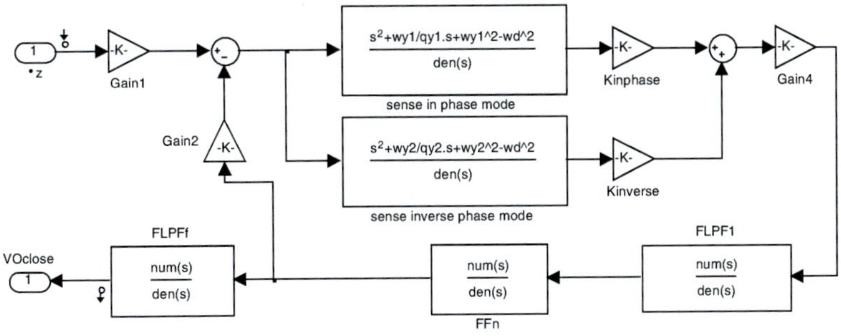

Fig. 5.19 Simulation diagram of sensing closed loop system

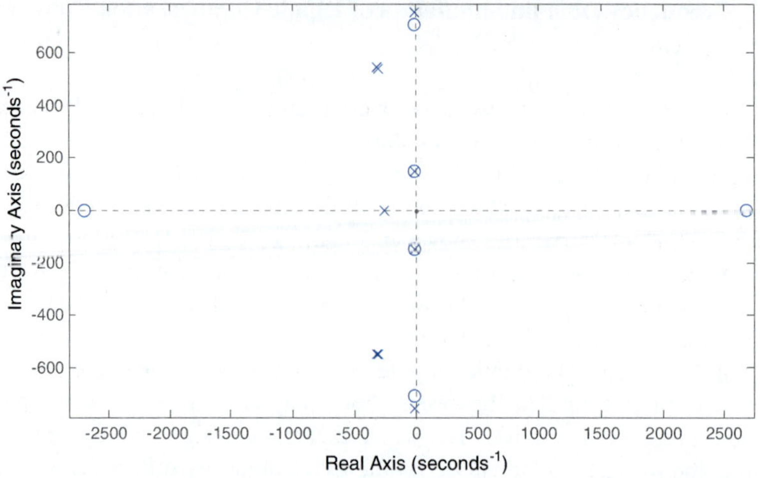

Fig. 5.20 Zero-pole distribution of sensing closed loop system (low frequency band)

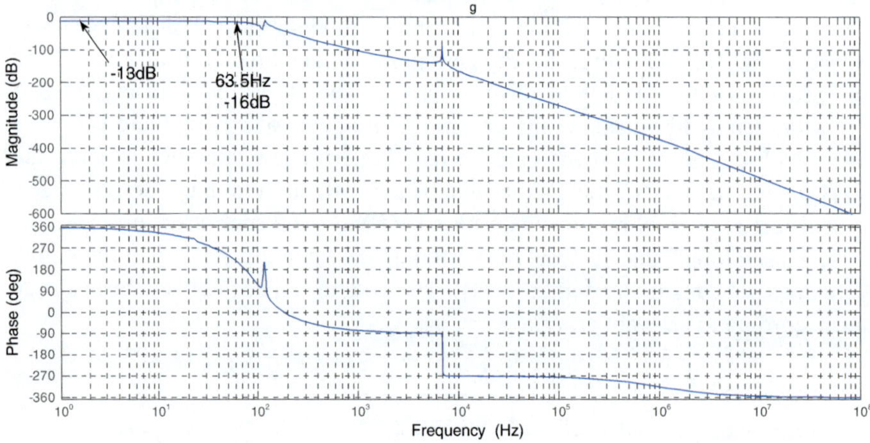

Fig. 5.21 Bod diagram of sensing closed loop system

5.6.1.5 Influence of Temperature on Dipole Compensation Controller

It can be seen from Figs. 2.9, 2.10, 2.11 and 2.12 in Chap. 2 that although the silicon-based MEMS gyroscope structure is in vacuum packaging, its modal quality factor will change greatly with the change of temperature (mainly because the Brownian motion of air molecules in the packaging will be caused by the wide range of temperature change). At the same time, the resonant frequency and $\Delta\omega_2$ of the gyro sensing mode also change with temperature. In combination with Formula (5.29), it can be seen that the dominant pole in the sensing closed loop system changes greatly with temperature. Table 5.4 shows the natural frequencies and quality factors of the

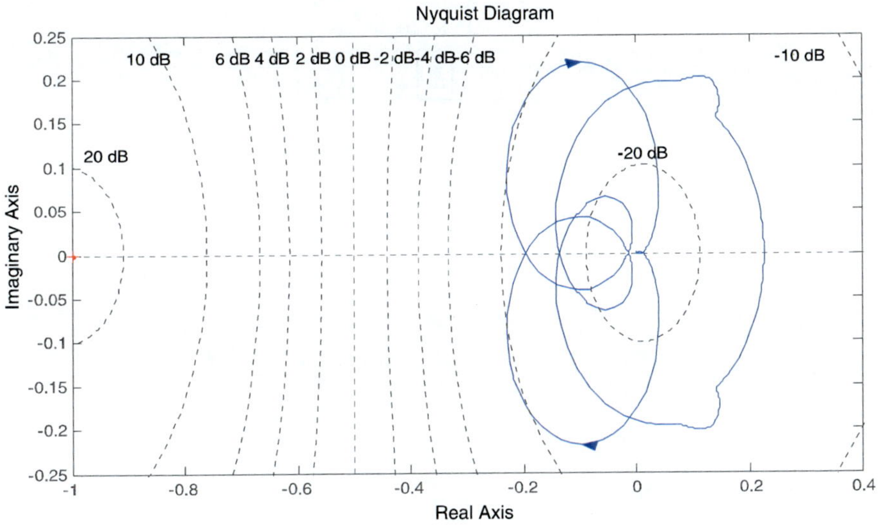

Fig. 5.22 Nyquist curve of sensing closed loop system

driving mode and sensing anti-phase mode of GY-027 gyro in the full temperature range under quadrature nulling.

Since the bandwidth expansion method in this book is based on the dipole compensation principle, the drift of the closed loop dominant pole with temperature will lead to the mismatch between the controller zero and the dominant pole, which will affect the dynamic characteristics of the closed loop system (and even lead to the instability of the closed loop system).It can be seen from the above table that the variation of $\Delta\omega_2$ is about 4.5% of normal temperature, while the total temperature variation of ω_{Y2}/Q_{y2} even exceeds 30% of normal temperature value. When the controller

Table 5.4 Variation of mechanical characteristic parameters of GY-027 structure at full temperature

Temperature(°C)	$\omega_d/2\pi$ (Hz)	$\omega_{y2}/2\pi$ (Hz)	$\Delta\omega_2/2\pi$ (Hz)	Q_x	Q_{y2}
-40	3494.7	3470.8	23.9	1773	1559
-30	3494.2	3470.1	24.1	1707	1495
-20	3493.1	3468.9	24.2	1626	1426
-10	3492.1	3467.9	24.2	1553	1362
0	3491.5	3467.0	24.5	1502	1315
10	3490.1	3465.5	24.6	1454	1275
20	3488.9	3464.1	24.8	1395	1224
30	3488.4	3463.6	24.8	1367	1199
40	3487.5	3462.6	24.9	1315	1151
50	3486.2	3461.3	24.9	1290	1130
60	3485.7	3460.7	25.0	1278	1122

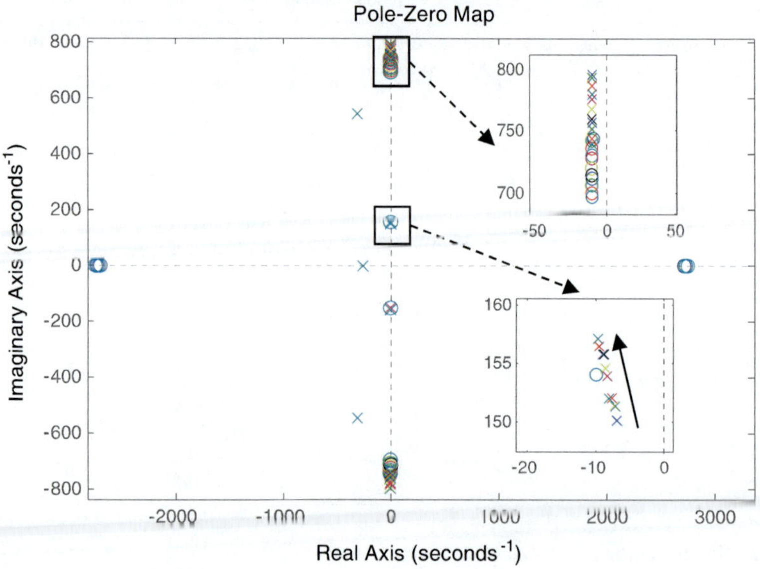

Fig. 5.23 Distribution of zero and pole of sensing closed loop system at different temperatures

parameters are at room temperature, the zero-pole and pole-pole distributions of the system are shown in Fig. 5.23 when the temperature changes from −40 to 60° C every 10° C.

The figure above mainly shows the distribution of low frequency zero and pole closely related to the system stability, and some regions with concentrated zero and pole are amplified. The enlarged figure on the upper right is the zero and pole corresponding to point B and point C in Fig. 5.5, and the enlarged figure on the lower right corresponds to point A. It can be seen that these poles are evenly distributed on the negative real axis, so the system has been in A stable state. The black solid arrow in the enlarged figure at the lower right corner indicates the direction of pole A as the temperature rises. It can be seen that when the temperature is lowest or highest, the compensation zero is furthest from pole A, and their dipole action is weakest. The structure of the above effect is that the peak generated by the A pole will gradually become prominent and deteriorate the in-band flatness of the silicon-based MEMS gyroscope, so that the +3 dB line limit bandwidth is exceeded in advance. In order to more intuitively reflect the bandwidth characteristics of gyro, the Bode diagram of the closed-loop system at each temperature point is shown in the Fig. 5.24.

In the figure, the frequency band near point A is locally amplified. Combining with Fig. 5.23, it can be seen that the curve corresponding to the pole closer to the compensation zero is flatter and its phase change is smaller. Extract the important parameters of curves at different temperatures in the figure above, as shown in Table 5.5. The index of "in-band flatness" in the table is defined as (lowest in-band gain—highest in-band gain)/(DC gain). According to the analysis of the data in Table 5.5,

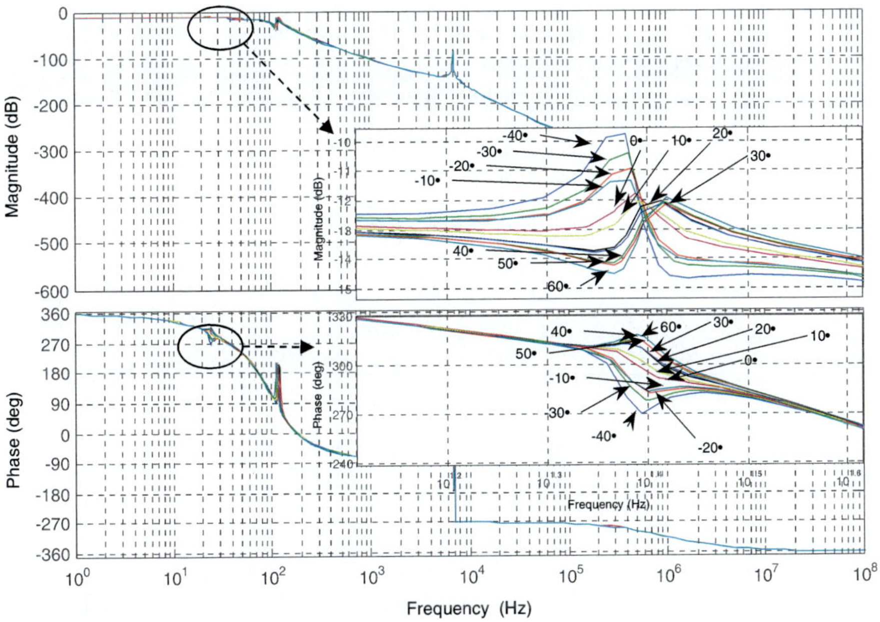

Fig. 5.24 Bode diagram of sensing closed loop system at different temperatures

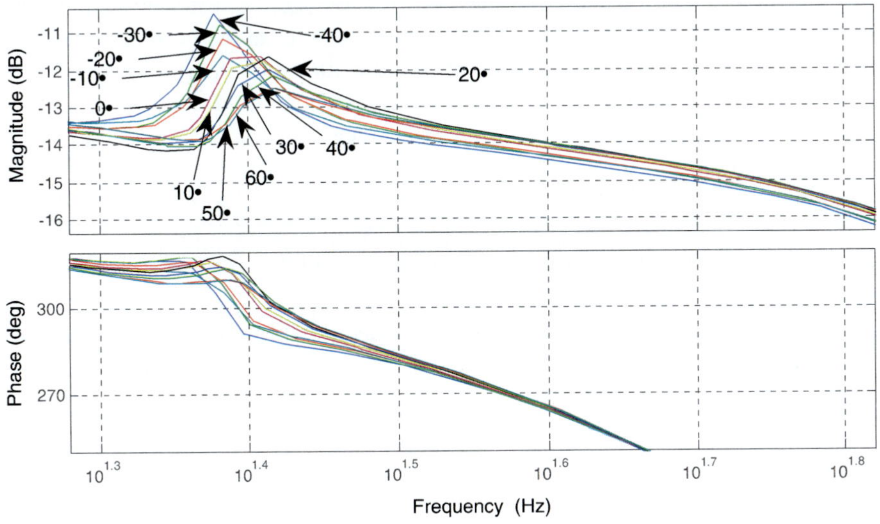

Fig. 5.25 Bod diagram of sensing closed loop system after compensation

Table 5.5 Simulated Bode diagram parameters of GY-027 under various temperature conditions after adopting dipole compensation controller

Temperature (°C)	DC gain (dB)	Highest frequency in the band (Hz)	Highest point gain in band (dB)	Lowest frequency in the band (Hz)	Lowest point gain in band (dB)	In-band flatness	Bandwidth (Hz)
−40	−12.38	23.9	−9.77	52.8	−15.38	0.45	52.8
−30	−12.46	24.1	−10.41	55.9	−15.46	0.41	55.9
−20	−12.50	24.2	−10.95	57.7	−15.5	0.36	57.7
−10	−12.50	24.2	−11.34	57.7	−15.5	0.33	57.7
0	−12.61	24.5	−11.79	61.1	−15.61	0.30	61.1
10	−12.65	24.6	−12.03	62.2	−15.65	0.29	62.2
20	−12.72	26.0	−12.03	65.0	−15.72	0.29	65.0
30	−12.72	26.0	−12.14	65.1	−15.72	0.28	65.1
40	−12.76	26.1	−12.06	65.2	−15.76	0.29	65.2
50	−12.76	26.1	−12.16	65.6	−15.76	0.28	65.6
60	−12.79	26.2	−11.91	66.1	−15.79	0.30	66.1

the compensation zero of the controller can well reduce the resonant peak generated by pole A within the full temperature range and expand the bandwidth beyond the resonant peak frequency. However, the bandwidth parameters and in-band flatness vary greatly, so temperature compensation can be carried out for the sensing closed loop controller to improve the dynamic characteristics of gyroscope in the full temperature range (Fig. 5.25).

5.6.1.6 Dipole Compensation Controller Temperature Compensation

Combining Eq. (5.29) and Fig. 5.23, it can be seen that the mode of the real part of the pole is much smaller than the imaginary part, so the focus of compensation should be the imaginary part of $P_{5,6}$ $\Delta\omega_2$.Meanwhile, according to Eq. (5.36), R_{f1} value is related to it, and compensation for R_{f1} does not affect other factors in Eq. (5.36).Thermistor compensation method is adopted, which has good repeatability, high reliability and easy circuit implementation [149]. As the temperature rises, the gyro frequency difference $\Delta\omega_2$ increases, the corresponding R_{f1} value should decrease. According to the R_{f1} expression, R_{f1b} can be set as A thermistor (see Table 5.6 for its value), and the closed loop Bode graph of the system at various temperatures can be obtained by matching the corresponding value of R_{f1}. For convenient observation, the value near the peak value of point A (the segment is the least flat) is shown in the figure below. See Table 5.6 for the main feature points in the figure. R_{f1a} and R_{f1c} are 4 and 1.8 KΩ, respectively. By comparing Tables 5.5 and 5.6, it is found that the bandwidth parameters of the full temperature are significantly improved after compensation, and the difference is basically within 2.5 Hz. At the same time, the flatness of the band is also improved to some extent compared with that before compensation, which proves the effectiveness of compensation. However, this method is limited to the design of individual header and lacks of generality, so it is of little significance in practical engineering application, and a new method needs to be developed.

5.6.2 Sensing Mode Closed Loop Controller Design Based on Proportional Integral Phase Lead Controlling Method

Although the dipole compensation controller introduced in the previous section can well compensate the in-band resonance peak caused by modal frequency difference and achieve good in-band flatness, its application conditions are harsh: Need for frequency and quality factor of sensing mode exact match, leading to the poor universality, only on a single silicon-based MEMS gyroscope design cannot satisfy the needs of the gyro rapid production in large quantities, this section studies based on the proportional integral and phase advance sensing closed loop nulling (PIPLC) controller can satisfy the general requirements of closed loop controller of silicon-based MEMS gyroscope in large quantities.

5.6.2.1 Working Principle of PIPLC Controller

According to the closed loop stability principle, the open loop must have sufficient amplitude and phase margin to ensure the closed loop stability, as shown in Fig. 5.26 (generally, the phase margin PM should be between 30° and 60°, and the amplitude margin GM should be greater than 6 dB).It can be seen from Figs. 5.5 and 5.8 that the resonant peak at point A is caused by two conjugate poles P_5 and P_6, and there will be A 180° phase change around them (the phase lag of each pole is 90°). At the same time, the higher quality factor of the silicon-based MEMS gyroscope sensing mode leads to A drastic phase change. Therefore, phase and amplitude must be corrected and compensated to ensure that the system has enough phase and amplitude stability margin [150].

5.6.2.2 PIPLC Controller Design

The design goal of PIPLC controller is to improve the system bandwidth on the premise of ensuring sufficient amplitude and phase margin. In addition, it can be seen from Fig. 5.12 that the system bandwidth and stability are only related to the middle closed loop, which can be expressed as a unit negative feedback system,

Table 5.6 Simulation baud diagram parameters of GY-027 under various temperature conditions after temperature compensation of dipole compensation controller

temperature (°C)	R_{f1b} resistance value (Ω)	R_{f1} equivalent resistance (Ω)	DC gain (dB)	High frequency (Hz)	High gain (dB)	Low frequency (Hz)	Low point gain (dB)	In-band flatness	Bandwidth (Hz)
−40	844	1486	−12.92	23.9	−10.48	62.8	−15.92	0.34	62.8
−30	883	1475	−12.93	24.1	−10.75	63.8	−15.93	0.40	63.8
−20	922	1463	−12.89	24.2	−11.14	63.7	−15.89	0.37	63.7
−10	961	1451	−12.82	24.2	−11.57	62.7	−15.82	0.33	62.7
0	1000	1440	−12.86	25.6	−11.62	64.8	−15.86	0.33	64.8
10	1039	1429	−12.83	25.8	−11.74	64.7	−15.83	0.32	64.7
20	1078	1418	−12.84	26.0	−11.62	65.2	−15.84	0.33	65.2
30	1117	1407	−12.76	26.0	−11.98	64.9	−15.76	0.30	64.9
40	1156	1396	−12.73	26.1	−12.16	65.0	−15.73	0.28	65.0
50	1195	1386	−12.67	26.1	−12.48	64.1	−15.67	0.25	64.1
60	1234	1376	−12.64	26.2	−12.45	64.1	−15.64	0.25	64.1

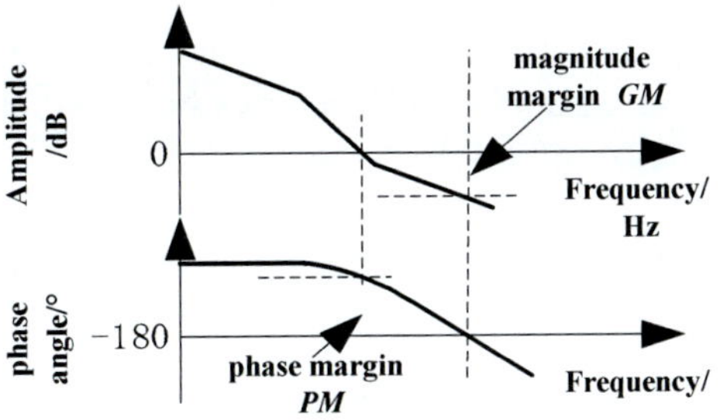

Fig. 5.26 Margin of a stabilized system

as shown in Fig. 5.27. Then, the open loop transfer function of the system can be expressed as:

$$H_{pi}(s) = \frac{V_{dac}K_{FBy}V_{dac}K_{yc}K_{pre}K_{amp}F_{LPF1}(s)F_{Fn}(s)G_{equal}(s)}{4m_y} \tag{5.38}$$

By comparing the above equation with Eq. (5.10), it can be seen that the frequency characteristics of other parts of the above equation, except the feedback controller FFn, can refer to the system bod diagram for detecting the open loop state. In general, the gain of low frequency band of the open-loop characteristics of the system should meet the requirement of steady-state error, the slope of middle frequency band (shear rate, passing through 0 dB line) should be −20 dB/dec, and the high frequency band with the required shear frequency ω_{cut} should decay as fast as possible to reduce the influence of high frequency noise on the system. At the same time, the series phase advance nulling device can increase the phase margin of the system, reduce the overshoot of the system response, increase the system bandwidth and speed up the response of the system. According to the above analysis, FFn adopts the form of first-order pure integral in low frequency band to minimize the steady-state error. In the middle frequency band, due to the phase lag of 180° near the frequency difference $\Delta\omega_2$, and the change is severe, seriously affecting the phase margin, so before $\Delta\omega_2$, two stage first-order PD link must be used to compensate the phase, then the slope after $\Delta\omega 2$ is −20 dB/ dec. It should be considered that point B ($\Delta\omega_B$) in Fig. 5.5 is the superposition of two zeros. After passing this point, the phase advance is 180° and the slope becomes +20 dB/ dec, so ω_{cut} should be in front of $\Delta\omega_B$. In the high frequency band, the slope returns to −20 dB/dec after passing point C, and then becomes −60 dB/dec under the influence of second-order low-pass filter FLPF1, which can already meet the demand of attenuating high frequency noise. However,

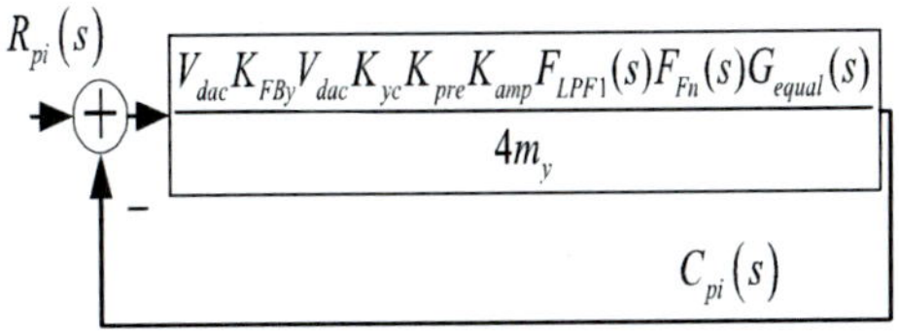

Fig. 5.27 Simplified block diagram of the closed loop

in order to match the PD link in the middle frequency band, the inertia link should be added in the high frequency band. At this time, the slope in the high frequency band is −80 dB/dec. It can attenuate high frequency noise very well. Is:

$$F_{Fn}(s) = k_{pi}\frac{s+\omega_{pi1}}{s}\frac{s+\omega_{pi1}}{s+\omega_{pi2}} \tag{5.39}$$

According to the above analysis and the optimization of the above formula parameters through simulation, $\omega_{Pi1} = 10\pi$ rad/s, $\omega_{Pi2} = 400\pi$ rad/s, $K_{pi} = 32$ (too large K_{pi} will destroy the stability of the system). The controller circuit is designed according to the above parameters, as shown in Fig. 5.28. In order to simplify the circuit and facilitate debugging, the controller consists of two stages of operational amplifier. The first stage can be regarded as PI controller, and the second stage is phase advance nulling device.

$$F_{Fn}(s) = \frac{V_{pio}}{V_{pii}} = \frac{R_{pi2}R_{pi4}}{R_{pi1}R_{pi3}}\frac{\left(s+\frac{1}{R_{pi2}C_{pi1}}\right)}{s}\frac{\left(s+\frac{1}{R_{pi4}C_{pi3}}\right)}{\left(s+\frac{1}{R_{pi3}C_{pi2}}\right)} \tag{5.40}$$

According to the parameters in Table 5.7, the simulation is performed in Multisim and compared with its system transmission, as shown in Fig. 5.29. In the figure, the circuit simulation results are basically consistent with the system simulation results.

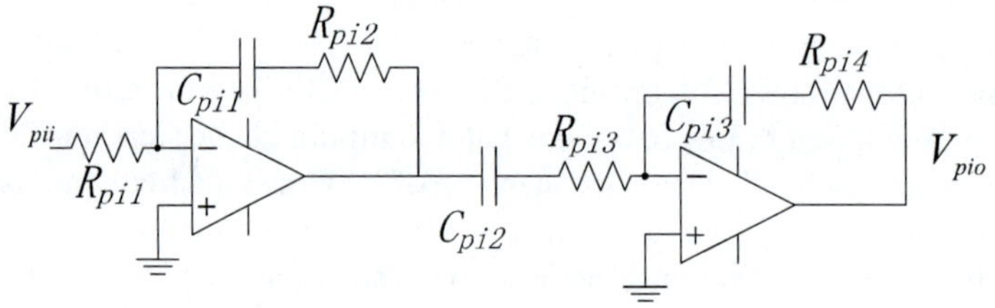

Fig. 5.28 PIPLC controller circuit

Table 5.7 *PIPLC* actual resistance and capacity parameters of the controller

Parameter label	Parameter value	Parameter label	Parameter value
R_{pi1}	130 kΩ	C_{pi1}	0.33 uF
R_{pi2}	100 kΩ	C_{pi2}	0.33 uF
R_{pi3}	2400 Ω	C_{pi3}	0.33 uF
R_{pi4}	100 kΩ		

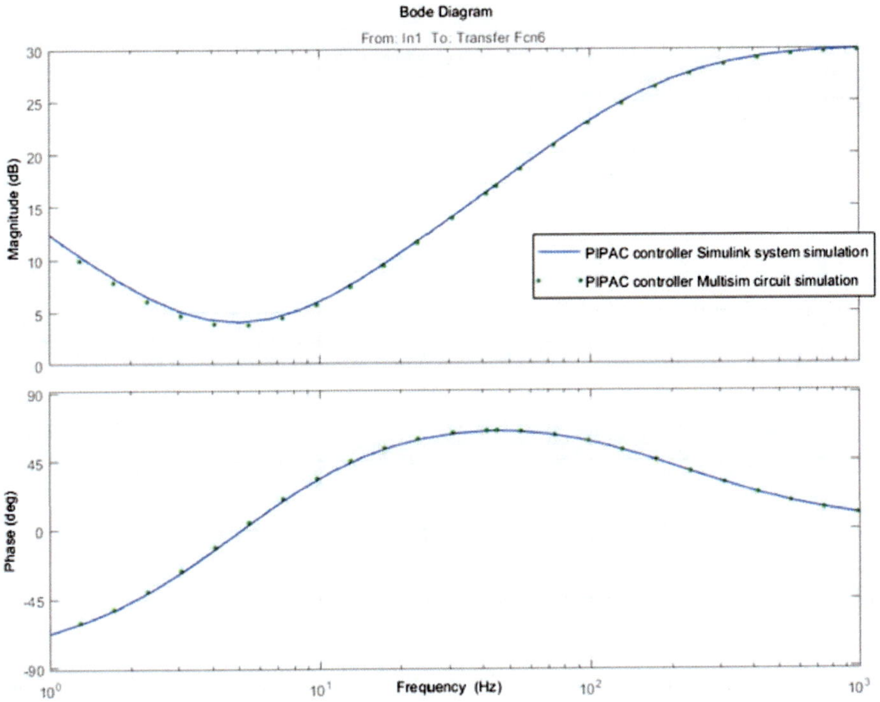

Fig. 5.29 PIPLC controller circuit simulation

5.6.2.3 PIPLC Controller Time Domain Analysis

The PIPLC controller is added on the basis of the previous model. The system works at t = 0 s, and the step signal of $\Omega_z = 100°/s$ is input at 3S to sense the sensing mode displacement signal and gyro output signal. The simulation curve is shown in Fig. 5.30. In the figure, the gyro reaches a stable state in 0.5 s. When the angular rate is input, the system also reaches a stable output state in a relatively short time, and the overshoot is very small. When the sensing displacement signal is amplified at constant angular rate input, it can be seen that the peak-to-peak value of the detected displacement is within 10–10 m, which is two orders of magnitude smaller than the displacement in the open loop state, indicating that the feedback force can basically balance the action of Coriolis force on the mass block, so that the mass block will not have large displacement in the sensing mode direction.

5.6.2.4 Frequency Domain Analysis of PIPLC Controller

Figures 5.31 and 5.32 respectively show the zero-pole and pole-pole distributions and Nyquist curves of the sensing PIPLC closed loop system. According to their respective criteria, the system is in a stable state. Figures 5.33 and 5.34 show the open loop

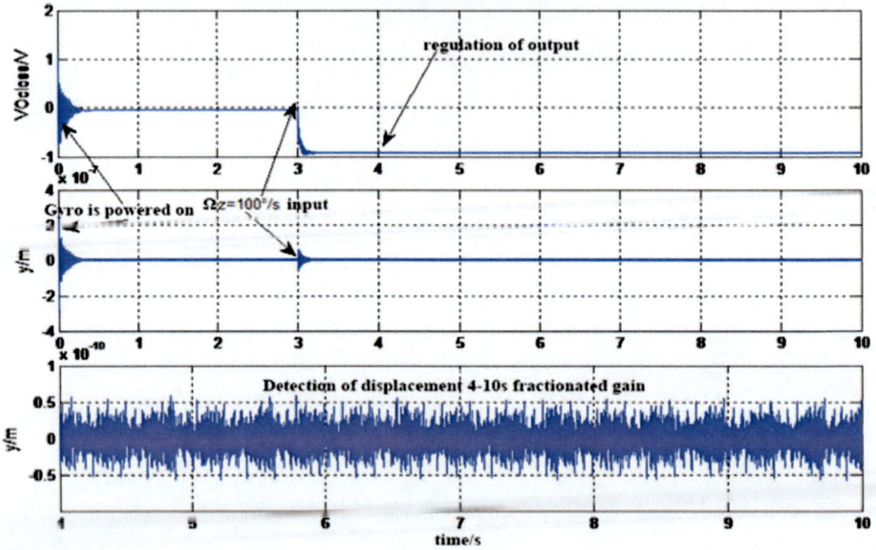

Fig. 5.30 PIPLC control system time domain simulation curve

and closed loop frequency characteristics of the sensing closed loop system. The minimum phase margin shown in Fig. 5.33 is 34.6° and the amplitude margin is 7.21 dB, both of which meet the indicators mentioned above. Figure 5.34 shows that the lowest inflection point value in the band is -13.8 dB and the highest inflection point is −10.4 dB. The bandwidth of the system is 100 Hz when the first two points do not exceed the ±3 dB range of −12.3 dB for DC. In addition, the figure also reflects that the bandwidth is limited to $\Delta\omega_B$, if you want to widen the bandwidth further you have to widen the difference between ω_{y1} and ω_{y2}.

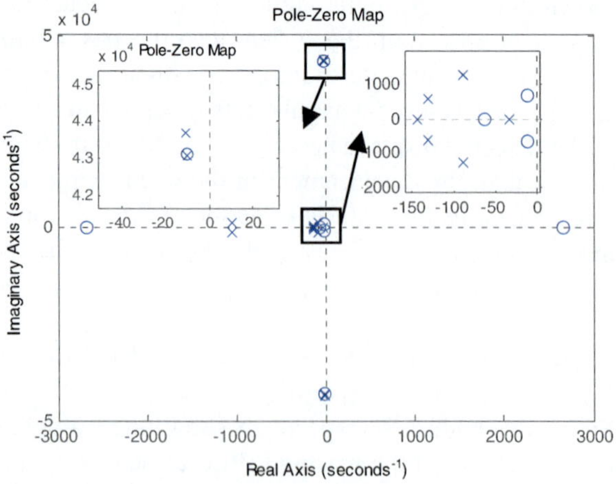

Fig. 5.31 PIPLC closed loop system zero-pole sensing diagram

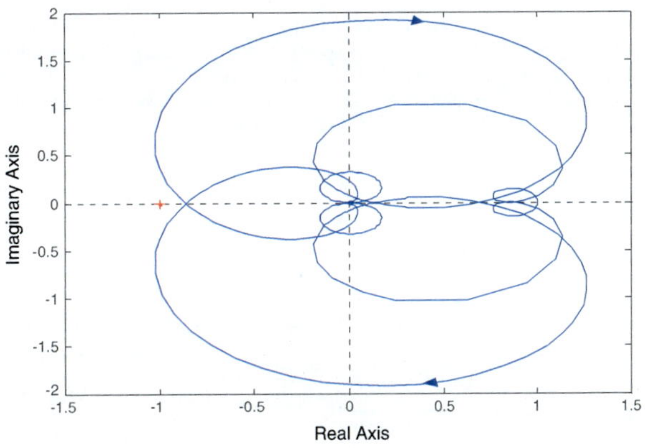

Fig. 5.32 Nyquist diagram for sensing PIPLC closed loop system

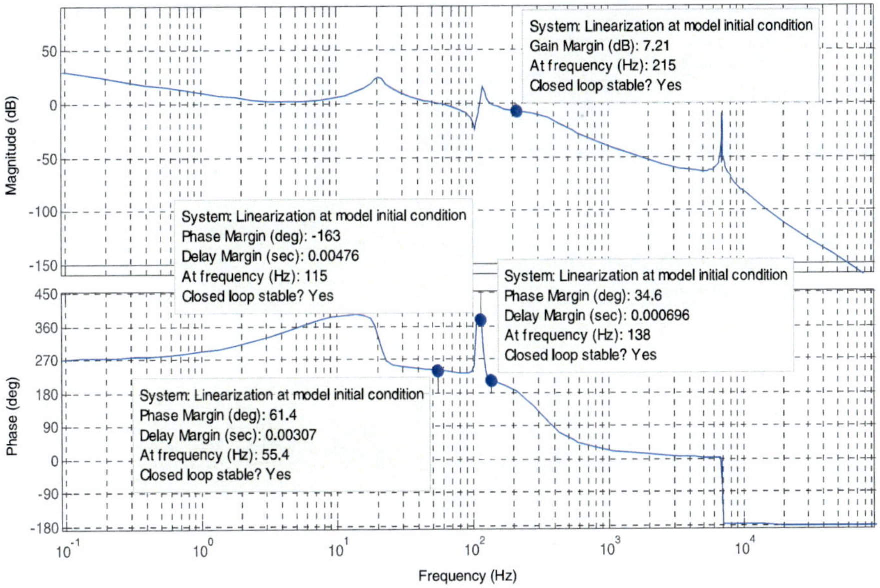

Fig. 5.33 Sensing open loop Bode diagram of the PIPLC system

5.6.2.5 Influence of Temperature on PIPLC

As can be seen from Table 5.4, the dual-mass linear vibration silicon-based MEMS gyroscope resonance frequency and quality factor within the full temperature range will produce a certain drift, but the change of $\Delta\omega2$ is less than 1 Hz. The controller parameters in Sect. 5.6.2.3 are adopted to substitute the mechanical parameters of each temperature point into the system to draw the sensing closed loop zero-pole diagram, as shown in Fig. 5.35. The system at each temperature point is in a stable state. Temperature change does not affect the stability of PIPLC control system.

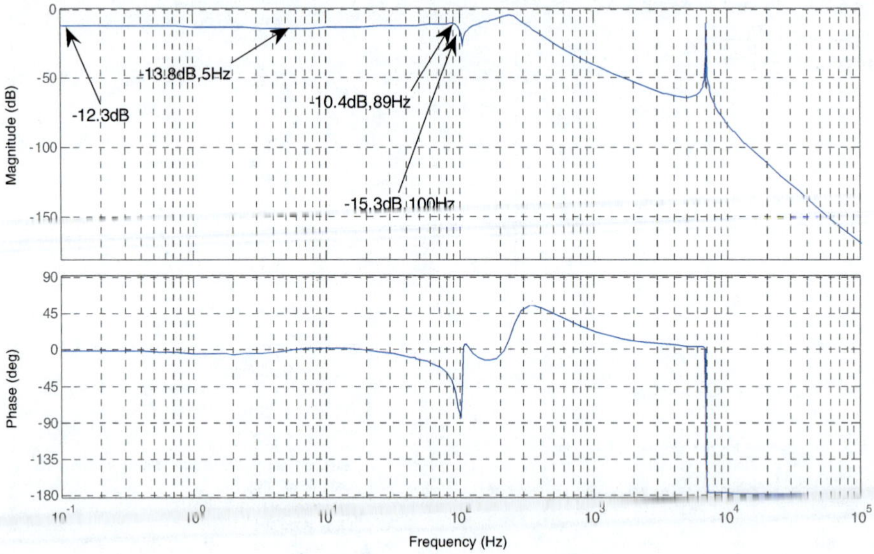

Fig. 5.34 Sensing closed loop Bode diagram of the PIPLC system

Fig. 5.35 Closed loop zero-pole diagram of PIPLC system at each temperature point

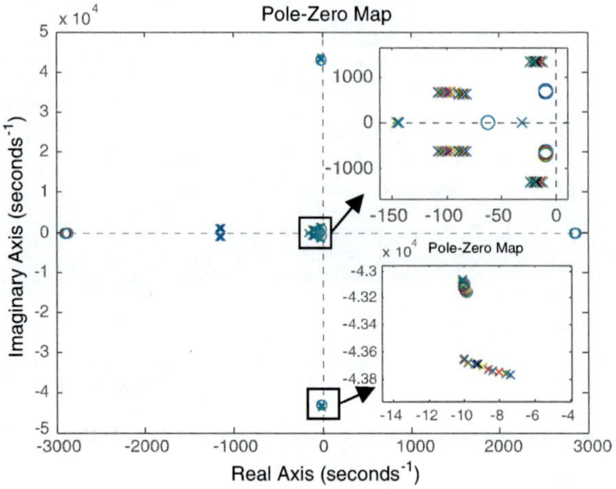

Figure 5.36 is the Bode diagram of the closed loop system under various temperature conditions. It can be seen that the curves basically overlap. Table 5.8 is the data of key points. In addition, by comparing the above table with Tables 5.5 and 5.6, it can be seen that the PIPLC controller has wider bandwidth and better in-band flatness than the dipole compensation controller, so the former has obvious advantages. This book chooses PIPLC as the object for relevant experiments.

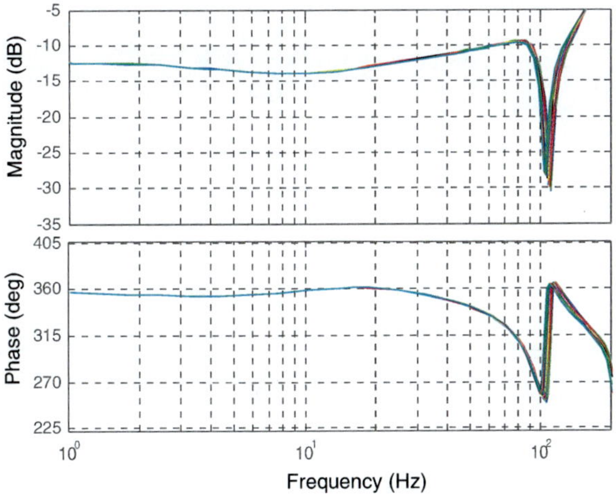

Fig. 5.36 Closed loop Bode diagram of PIPLC system at each temperature point

Table 5.8 Simulated Bode diagram parameters of GY-027 under various temperature conditions after PIPLC controller is adopted

Temperature (°C)	DC gain (dB)	Lowest inflection frequency in the band (Hz)	Lowest inflection gain in band (dB)	Frequency of highest inflection point in the band (Hz)	Maximum inflection gain in band (dB)	In-band flatness	Bandwidth (Hz)
−40	−12.393	7.80	−13.890	84.0	−9.449	0.36	103
−30	−12.395	7.79	−13.905	84.2	−9.450	0.36	103
−20	−12.397	7.79	−13.915	84.1	−9.471	0.36	102
−10	−12.401	7.79	−13.916	83.8	−9.512	0.36	101
0	−12.403	7.77	−13.940	84.1	−9.511	0.36	101
10	−12.407	7.76	−13.950	83.6	−9.533	0.36	101
20	−12.410	7.75	−13.965	82.8	−9.550	0.36	100
30	−12.412	7.75	−13.965	82.4	−9.561	0.35	99.7
40	−12.414	7.75	−13.974	81.8	−9.592	0.35	99.2
50	−12.417	7.75	−13.976	80.9	−9.621	0.35	98.7
60	−12.418	7.75	−13.985	80.5	−9.628	0.35	98.2

5.6.2.6 Closed Loop Experiment of PIPLC Controller

According to the above analysis and circuit parameters, the circuit is built on PCB board, and the closed loop bandwidth, the sensing mode displacement in open loop and closed loop state, and the step input characteristics of the system are actually measured by the sensing force feedback comb excitation method. Figure 5.37 shows the simulation and measured curve of unit negative feedback system bandwidth as

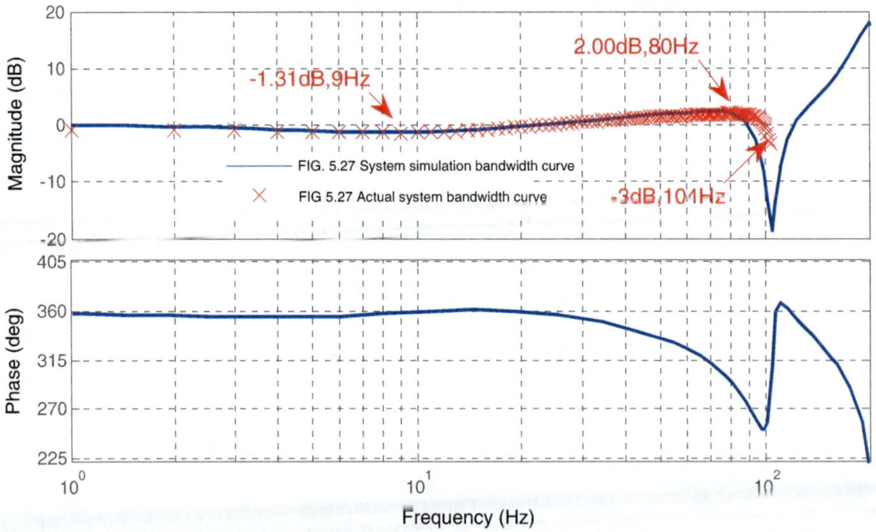

Fig. 5.37 PIPLC closed loop control system bandwidth measurement curve

shown in Fig. 5.27. After ignoring the influence of output stage low-pass filter, this curve is the effective bandwidth of gyro. For convenience of observation, only 1–200 Hz is selected in the figure. It can be seen from the figure that the frequency corresponding to the −3 dB point is 104 Hz, the frequency at the lowest inflection point in the bandwidth is 9 Hz and the amplitude is −1.31 dB, and the highest point is 2 dB corresponding to 80 Hz. The limiting factor of bandwidth is $\Delta\omega_B$, which is consistent with theoretical analysis.

Step response of the closed loop system was tested. After the system was powered on and worked in a fixed state, the step signal was used to input voltage on the sensing force feedback comb to simulate the input of angular rate $\omega_z = 100°/s$. The corresponding step diagram of the system was obtained, as shown in Fig. 5.38. Figure in the tope curve for step input curve V_{ftest}, bottom for sensing mode displacement signals, known from the figure sensing mode displacement within 120 ms to reach equilibrium, and Fig. 5.30 coincided basically with the simulation results, even in peak signal input system and only about 50 mv, by further optimization of system parameters of the peak can be better inhibition. Figs. 5.39 and 5.40 respectively show the sensing mode displacement curves when input $\omega_z = 100°/s$ equivalent angular rate is sensed in the open loop and closed loop states. The medium reference curve in the figure is the driving mode displacement curve. It can be seen that the peak-to-peak value of sensing mode displacement signal in the open loop state is greater than 700 mV, while the peak-to-peak value after closed loop is about 10 mV. And it is basically the noise term without Coriolis signal characteristics. The results are consistent with those in Fig. 5.30.

In order to verify the full-temperature characteristics of the PIPLC controller, the bandwidth of the closed loop system was tested at −40 and 60 °C. The test results

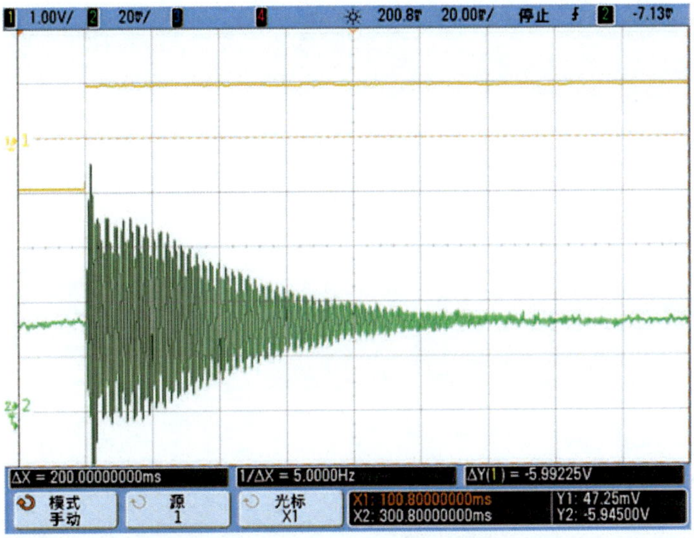

Fig. 5.38 Step response experiment of PIPLC closed loop control system

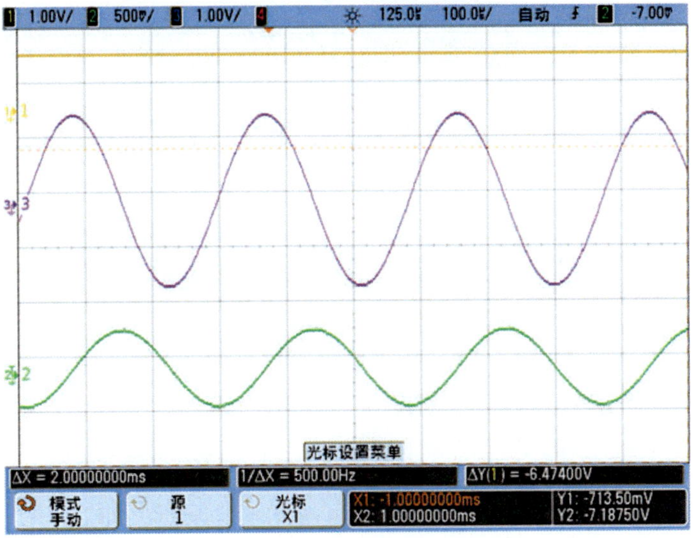

Fig. 5.39 Sensing open loop state sensing mode displacement curve

are shown in Fig. 5.41. The dark and light solid lines in Fig. 5.36 are simulation curves at −40 and 60 °C respectively, and the dark "×" and light "O" are measured curve points. It can be seen from the figure that the system bandwidths of −40 and 60 °C are 106 and 98 Hz respectively, which is consistent with the simulation results in Table 5.8 (the higher the temperature, the narrower the bandwidths).

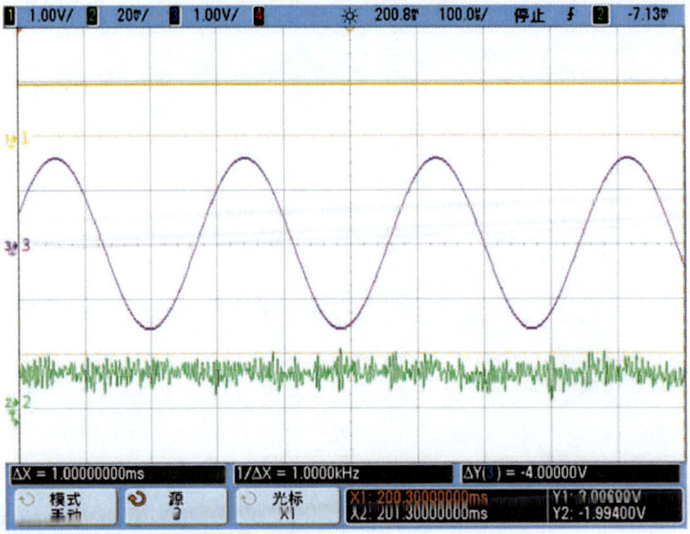

Fig. 5.40 Sensing mode displacement curve of sensing closed loop state

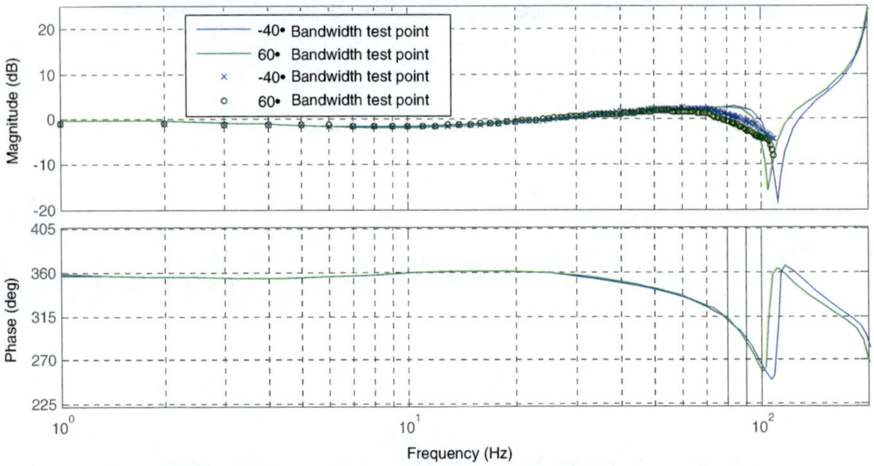

Fig. 5.41 −40 °C and 60 °C temperature bandwidth simulation and test curve

5.7 Dual-Mass Linear Vibration Silicon-Based MEMS Gyroscope Frequency Tuning Technology Research and Simulation

Dual-mass linear vibration silicon-based MEMS gyroscope frequency tuning technology can reduce the equivalent stiffness and natural frequency of the sensing mode by sensing the negative static force stiffness generated by the comb teeth of the mode pressure, so as to make the resonant frequency of the sensing and driving mode equal (mode matching state). The difficulty of this technology lies in the determination of the frequency difference between the two modes. According to Eq. (2.30), when modal matching is $\Delta\omega \approx 0$, the gyro obtains the maximum mechanical sensitivity, which is of great significance to improve the signal-to-noise ratio of the gyro output signal.

5.7.1 Frequency Tuning Principle Based on Driving Displacement Phase Information

According to the phase information in Eqs. (2.26) and (2.27), the driving displacement signal is $x(t) = A_x\cos(\omega dt)$, and when the frequency is tuned, φ_y is 90°, then the sensing displacement signal generated by Coriolis force is $y(t) = A_y\cos(\omega dt)$, which is in phase with x(t). Then the quadrature signal in the sensing loop should have a 90° phase difference with x(t) as $\sin(\omega_{dt})$ component. The relation curve between the composition of the quadrature signal in the sensing channel and the frequency difference Δf is shown in Fig. 5.42.

It can be seen from Fig 5.42 that the amplitude of the quadrature signal reaches its maximum when the frequency difference approaches 0, at which the $\cos(\omega_{dt})$ component is about 0. Meanwhile, Δf at the amplitude of the COS (ω_{dt}) component within the range of −1 to +1 Hz can be used as the criterion of modal tuning. Since the quadrature signal always exists in the structure, the phase difference between the quadrature signal and x(t) is adopted in this section as the basis for judging frequency tuning, so the quadrature signal should not be completely eliminated in the quadrature nulling link (only the quadrature signal with a small amplitude should

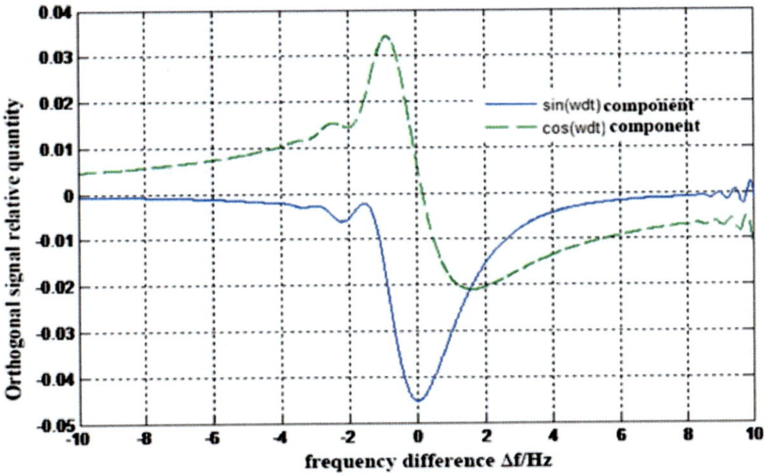

Fig. 5.42 Relation between quadrature signal components and frequency difference

be retained to reduce its influence on the Coriolis signal).It can be seen from Chap. 4 that the quadrature coupling stiffness method will affect the sensing mode resonance frequency, so in order to improve the stability of the system with frequency tuning links, the quadrature force nulling method should be used to weaken the quadrature signal. In addition, since the tuned comb can only reduce ω_{y2}, the design value of ω_{y2} should be ensured to be greater than ω_x. In this case, according to Eq. (5.10), it can be seen that the open loop system contains a pair of conjugate zeros, as shown in Eq. (5.41). The conjugate zeros will produce large depressions in the amplitude-frequency characteristic curve, directly limiting the system bandwidth. Figure 5.43 can be obtained by combining the principle of sensing in-phase and anti-phase mode superposition.

$$z_{t1,t2} = -\frac{\omega_{y2}}{2Q_{y2}} \pm \sqrt{\left(\omega_{y2}^2 - \omega_d^2\right) - \frac{\omega_{y2}^2}{4Q_{y2}^2}}\, j \tag{5.41}$$

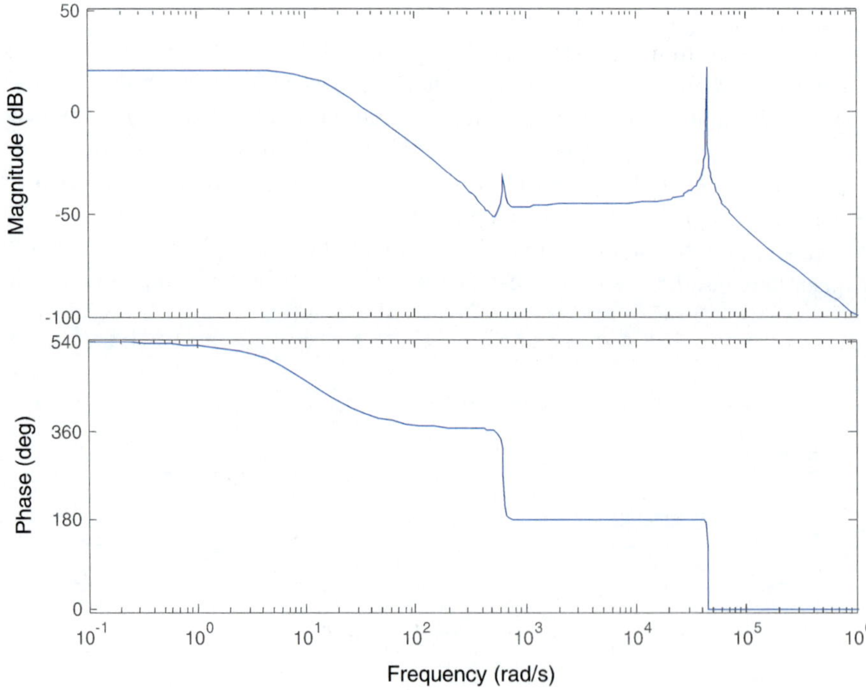

Fig. 5.43 Bode diagram of an open loop system in tuned state

5.7.2 *Dual-Mass Linear Vibration Silicon-Based MEMS Gyroscope Frequency Tuning Model*

According to the analysis of the above section, the tuning link and frequency difference solution module are added on the basis of the simulation model of Simulink system in Chap. 3, and $\omega_{y2} = \omega_x + 20\pi$ rad/s is set at the same time. Although the maximum mechanical sensitivity can be achieved when $\Delta\omega = 0$, $\Delta\omega \neq 0$ is usually set to sense the expansion of closed loop bandwidth. In this book, $\Delta\omega = 1$ Hz is selected. At the same time, it should meet the requirements of $\omega_{y2} < \omega_x$ to avoid the effect of depression on bandwidth in Fig. 5.43. Figure 5.44 shows the automatic frequency tuning model. The set voltage in the frequency tuning module can be set to achieve different steady state ΔF. Meanwhile, the negative stiffness conversion coefficient of electrostatic force can be adjusted by the amplification of applied voltage. Figure 5.45 is the simulation curve of the frequency automatic tuning system. It can be seen from the figure that the system reaches a stable state within 2 s, ΔF is controlled between 1.025 and 1.03 Hz, and the control purpose is achieved, and the system has no overshoot.

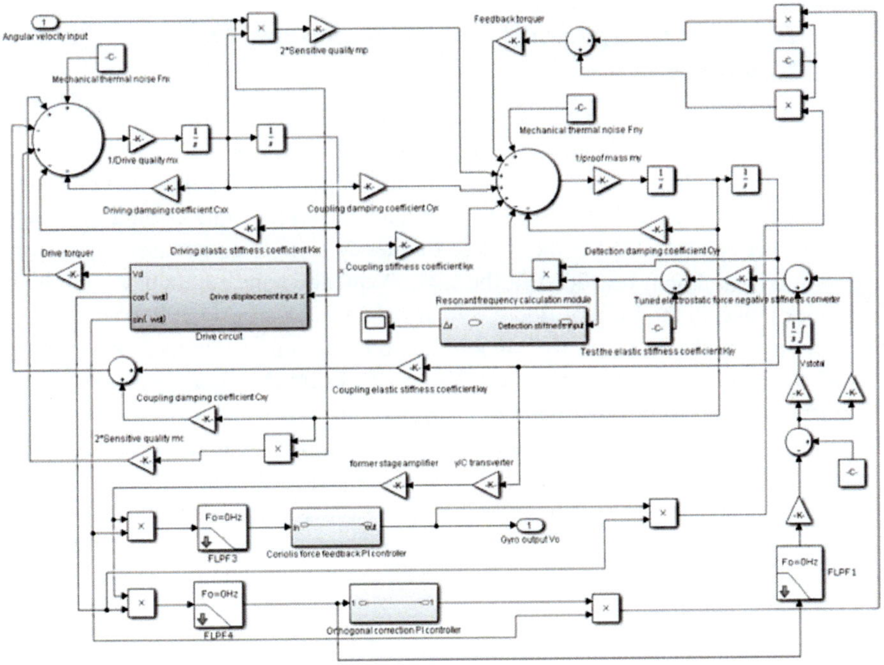

Fig. 5.44 Dual-mass linear vibration silicon-based MEMS gyroscope frequency automatic tuning system model

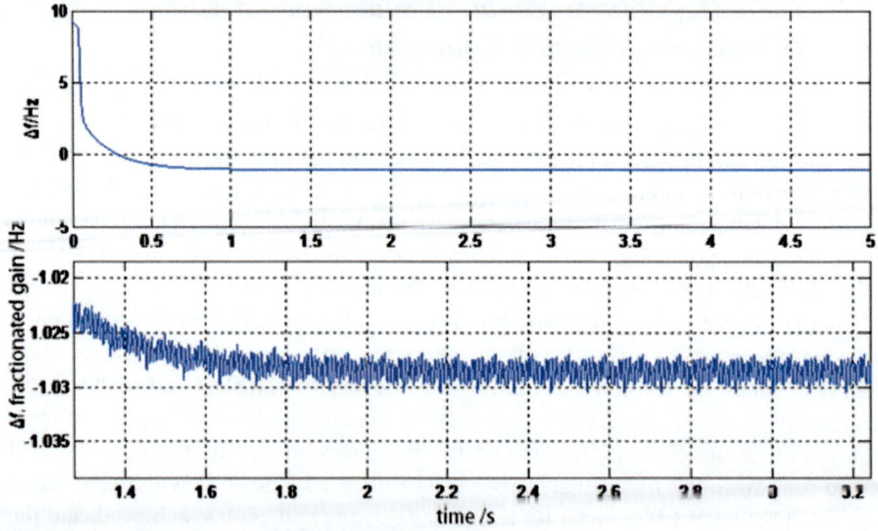

Fig. 5.45 Simulation curve of automatic frequency tuning system

5.7.3 Bandwidth Expansion Under Frequency Tuning Condition

In the analysis in the previous section, the steady state has $\omega_{y2} \approx \omega_x$-2π rad/s. According to the analysis in the previous section, a pure integral link can be added in the low frequency band of the system to reduce the steady state error, a two-stage PD controller can be added in the middle frequency band to compensate the phase, and an inertia link can be added in the high frequency band to attenuate the high frequency noise, in the form of Eq. (5.39). $\omega_{pi1} = \pi$ rad/s, $\omega_{pi2} = 200\pi$ rad/s, $K_{pi} = 50$, combined with Fig. 5.43, we can get the bod diagram of the open loop system (Fig. 5.46) and the closed loop system (Fig. 5.47). From the figure, we can see that the phase Angle margin and amplitude margin of the system are respectively 16 dB and 103°, and the system bandwidth is 82 Hz. In addition, point B in Fig. 5.5 still limits the system bandwidth. Figures 5.48 and 5.49 show the zero-pole distribution and Nyquist curve of the sensing closed-loop system at 1 Hz frequency difference, where all poles of the system are located at the negative half axis of the real axis, and the Nyquist curve does not contain (0, −1j) points, which proves the stability of the system.

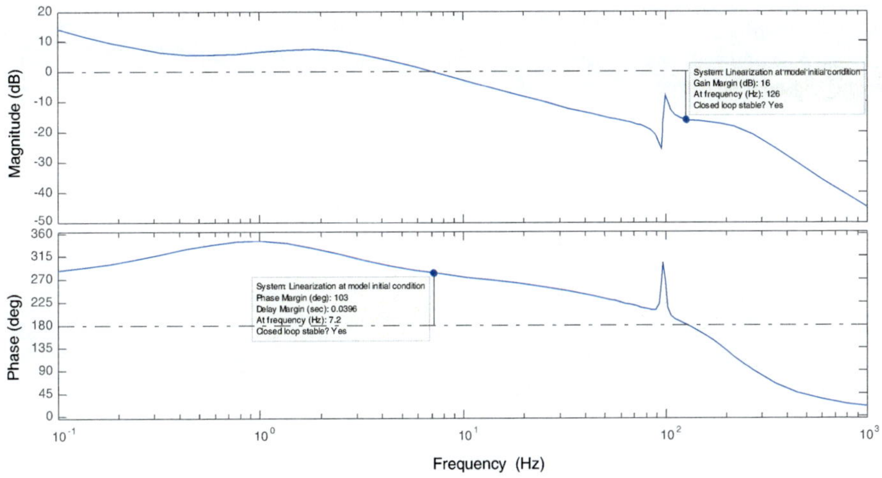

Fig. 5.46 Open loop Bode diagram of tuned state closed loop system

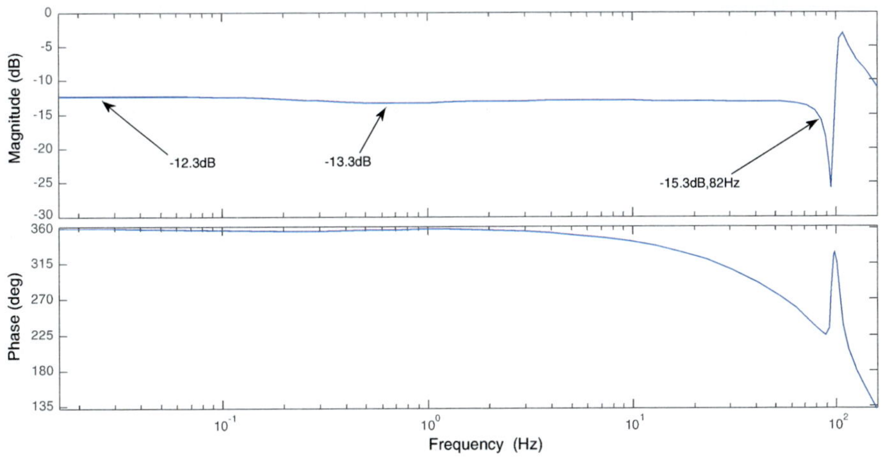

Fig. 5.47 Sensing closed loop Bode diagram of tuned state

5.8 Chapter Summary

In this chapter, the Coriolis sensing and frequency tuning techniques of dual-mass linear vibration silicon-based MEMS gyroscope are studied. In this paper, a force-feedback comb excitation method is proposed to simulate the input Angle rate with external signal source, and the sensing mode of the meter head under the interaction of in-phase and anti-phase modes is analyzed theoretically, and verified by force-feedback excitation method. The working principle of the open loop sensing loop

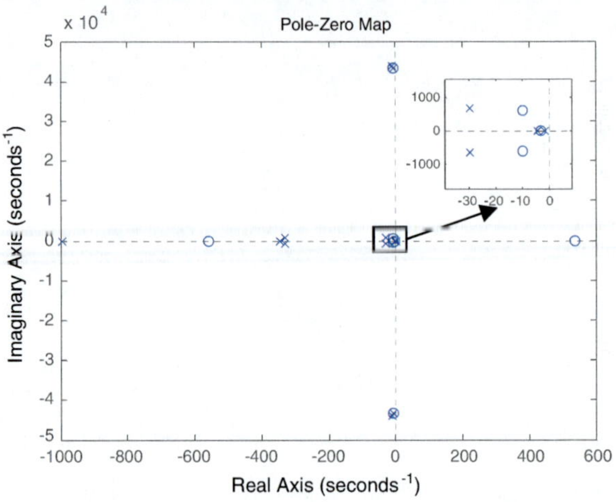

Fig. 5.48 Distribution of zero and pole of tuned state sensing closed loop system

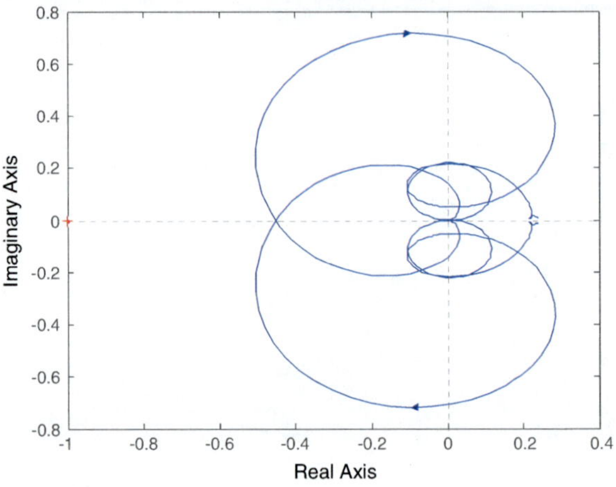

Fig. 5.49 Nyquist curve of tuned state sensing closed loop system

is analyzed, and the time-domain response and frequency-domain bandwidth characteristics of the open loop sensing loop are simulated and analyzed in Simulink simulation environment. It is pointed out that the mechanical frequency difference between the driving mode and sensing mode is the limiting factor of the bandwidth of the open loop system. The sensing closed loop controller is designed based on dipole compensation method and proportional integral phase zeroing method (PIPLC) respectively, and the respective model has carried on the simulation of time domain and frequency domain, and the simulation results were compared in view of the practical application requirement for controller, this chapter select PIPLC controller to a step input, bandwidth, zero bias stability experiment. The experimental results show that the PIPLC controller has good dynamic characteristics and the ability to expand the gyro test bandwidth. The experimental results also prove that the closed loop can effectively suppress the gyro sensing mode motion amplitude, which helps improve the dual-mass linear vibration silicon-based MEMS gyroscope anti-vibration and impact ability. At the same time, compared with the sensing open loop system, The closed loop system shows significant improvement in zero-bias stability, nonlinearity and asymmetry of scale factors, and temperature-dependent parameters (experimental results will be described in detail in Chap. 6). In addition, the closed loop is tested on the wide characteristics of the whole temperate zone, and the system can work stably in the full temperature range, and the higher the temperature, the narrower the bandwidth. This conclusion is consistent with the simulation results.

This chapter also studies the technique of automatic frequency tuning of silicon-based MEMS gyroscope, through the study found that when sensing anti-phase mode is higher than the driving anti-phase mode, the system of the amplitude frequency characteristics appeared directly caused by the conjugate zero huge sag limits the frequency of tuning state expansion, so the sensing anti-phase frequency should be adjusted to driving anti-phase made of the following in order to expand the system bandwidth. The frequency tuning module is introduced into the system model, and the simulation results show that the system can adjust the frequency difference of the driving and sensing modes to the set value in a short time (the set value in this chapter is 1 Hz). At the same time, a PIPLC sensing closed loop controller is designed for the tuning state to expand the silicon-based MEMS gyroscope test bandwidth, and the simulation results prove the feasibility of the theoretical research.

Chapter 6
Temperature Influence on Silicon-Based MEMS Gyroscope and Suppression Method

6.1 Chapter Introduction

Because the sensitive structure of silicon-based MEMS gyroscope is sensitive to temperature, the temperature has a great influence on the silicon-based MEMS gyroscope, and the influence not only occurs in the change of the working environment temperature of the gyroscope, but also the heat generated by the circuit of the silicon-based MEMS gyroscope when the environment temperature is constant is enough to make the output signal of the gyroscope drift. This chapter first analyzes the influence of temperature on stiffness and damping from the perspective of structure, and then summarizes the methods for inhibiting the influence of temperature on the silicon-based MEMS gyroscope itself from multiple perspectives.

6.2 Temperature Influence on Silicon-Based MEMS Gyroscope and Suppression Method

From the structural analysis in the previous chapters, the most direct parameters that affect the modal motion in the silicon-based MEMS gyroscope structure are the equivalent stiffness and quality factor of the modal. However, due to the influence of various factors in the working environment of the gyroscope, the stiffness coefficient and quality factor will change, and the influence of temperature is the most obvious.

6.2.1 Temperature Influence on Modal Stiffness Coefficient

In the dual-mass linear vibration silicon-based MEMS gyroscope structure of this paper, the driving and sensing modes adopt the same U-beam structure. Taking the driving beam as an example, as shown in Fig. 6.1, the two long beams are equal in

H. Cao, *Dual-Mass Linear Vibration Silicon-Based MEMS Gyroscope*,
https://doi.org/10.1007/978-981-19-9247-6_6

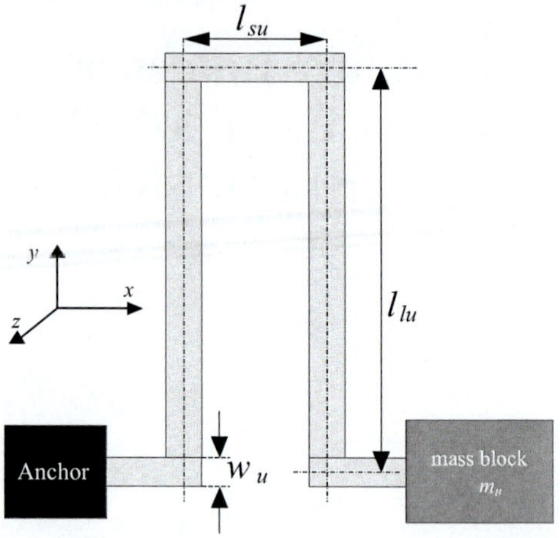

Fig. 6.1 U-beam system model

length. One end of the two long beams is connected to the two ends of the short beam, and the other end is connected to the anchor point and the driving mass. Then the stiffness of the U-shaped beam in the x direction is:

$$k_{xmu} = \frac{Eh_u w_u^3 (2l_{lu} + l_{su})}{8l_{lu}^3 l_{su} + 4l_{lu}^4} \tag{6.1}$$

In the formula, k_{xmu} is the stiffness of U-beam in x direction; E is the Young modulus of silicon material; h_u is the thickness of beam arm; w_u is the beam arm width; l_{lu} and l_{su} are the lengths of long beams and short beams, respectively. In practical engineering, U-shaped beams are often subjected to residual stresses, so the total stiffness of the model in the x direction in Fig. 6.1 can be expressed as:

$$k_{xm} = k_{xmu} + k_{xmr} = \frac{Eh_u w_u^3 (2l_{lu} + l_{su})}{8l_{lu}^3 l_{su} + 4l_{lu}^4} + \beta_x \sigma_R h_u w_u \tag{6.2}$$

In the formula, k_{xmr} is the equivalent stiffness of residual stress; β_x is a coefficient related to the vibration mode; σ_R is residual stress. From Figure are 2.5, it can be seen that in the driving mode, each driving equivalent mass block has six driving beams and two connecting beams. Because the two mass blocks are reverse motion, the equivalent stiffness of each connecting beam is equivalent to twice the driving beam, so the equivalent stiffness of the driving mode $k_x = 10k_{xm}$, the resonant frequency of the driving mode should be expressed as:

$$\omega_x = 2\pi f_x = \sqrt{\frac{k_x}{m_x}} = \sqrt{\frac{10}{m_x}\left(\frac{Eh_u w_u^3 (2l_{lu} + l_{su})}{8l_{lu}^3 l_{su} + 4l_{lu}^4} + \beta_x \sigma_R h_u w_u\right)} \tag{6.3}$$

In addition, under the influence of temperature, there are:

$$\Delta E = k_E \Delta T E_0 \tag{6.4}$$

$$\Delta l = \alpha_S \Delta T l_0 \tag{6.5}$$

$$\Delta \sigma_R = \frac{E_0}{1+\nu}(\alpha_S - \alpha_G)\Delta T \tag{6.6}$$

where ΔE, Δl and $\Delta \sigma_R$ are the variation of Young modulus, structural length and residual stress, respectively; ΔT is the variation of temperature relative to T_0; $k_E \approx -50$ ppm/°C is the temperature coefficient of Young modulus;$E_0 = 1.3 \times 10^{11}$ Pa and l_0 are the values of Young modulus and structural size at T_0 temperature, respectively; $\nu = 0.27$ is the Poisson's ratio of silicon material; $\alpha_G = 3.25 \times 10^{-6}$/°C is the thermal expansion coefficient of glass material; α_s is the thermal expansion coefficient of silicon material, and there is:

$$\alpha_S = \{3.725\left[1 - e^{-5.88\times 10^{-3}(T-124)}\right] + 5.548 \times 10^{-4}T\} \times 10^{-6} \tag{6.7}$$

In the formula, T is the Warsaw temperature. So the influence of temperature on elastic stiffness is mainly from four aspects: Young modulus, thermal expansion coefficient, residual stress and size, the change of stiffness coefficient can be expressed as:

$$\Delta k_x = \frac{\partial k_x}{\partial E}\Delta E + \frac{\partial k_x}{\partial \alpha_S}\Delta \alpha_S + \frac{\partial k_x}{\partial \sigma_R}\Delta \sigma_R + \frac{\partial k_x}{\partial l_{lu}}\Delta l_{lu} + \frac{\partial k_x}{\partial l_{su}}\Delta l_{su} + \frac{\partial k_x}{\partial h}\Delta h + \frac{\partial k_x}{\partial w_u}\Delta w_u \tag{6.8}$$

By substituting the corresponding design parameters of the dual-mass linear vibration silicon-based MEMS gyroscope structure in Table 2.1, the equivalent stiffness of the driving mode varies with temperature as follows:

$$\Delta k_x \approx -0.018\Delta T - 4.8 \times 10^{-5}\Delta T^2 \tag{6.9}$$

After ignoring the quadratic term of small temperature variation, the equivalent stiffness of the modal is inversely proportional to the temperature. Table 6.1 summarizes the variation of each part with temperature. The analysis shows that the change of Young modulus with temperature plays a key role in the temperature drift of equivalent stiffness. In addition, by increasing the structural thickness and beam width of the U-shaped beam, the trend of equivalent stiffness changing with temperature can be reduced, thereby reducing the temperature coefficient of resonant frequency.

Table 6.1 Effect of parameters on temperature coefficient of stiffness coefficient

Parameter	Temperature coefficient produced
Young's modulus E	$-0.023\ \Delta T$
Thermal expansion coefficient α_s	$0.0036\ \Delta T$
Residual stress σ_R	$-0.000048\ \Delta T^2$
Long driving (connection) beam l_{lu}	$-0.0035\ \Delta T$
Driving(connection) beam width w_d	$0.0012\ \Delta T$
Structural thickness b	$0.0036\ \Delta T$
Driving (connection) beam short beam l_{su}	$-0.00014\ \Delta T$

6.2.2 Temperature Influence on Modal Damping Coefficient

The damping in silicon-based MEMS gyroscope structure mainly includes two parts: mechanical damping and air damping. Under the condition of approximate vacuum (pressure less than 1 mTorr), mechanical damping is the main damping, which includes thermoelastic damping, electronic damping and anchor loss damping. However, in engineering applications, it is difficult for the silicon-based MEMS gyroscope structure package to reach this pressure level, even if it reaches it, it is also difficult to maintain (some external air will penetrate into the package after a period of time). Therefore, the pressure in the package is usually about 10 Torr magnitude even after vacuuming, and the air damping is still the dominant damping in this state. This section focuses on the model of air damping affected by temperature.

The air damping of the driving mode consists of two parts (taking the driving mode as an example, the sensing mode is the same): the slide damping of the driving frame and Coriolis mass with the glass substrate, and the damping between the driving comb. It is pointed out in the previous chapter that the comb teeth of the driving and sensing modes in the dual-mass linear vibration silicon-based MEMS gyroscope structure in this book are all sliding mode structures, and the damping model is shown in Fig. 6.2. When the comb moves relatively along the x-axis direction, the sliding film damping will be generated between the parallel voids of the comb, and the pressure film damping will be generated between the top and bottom ends of the comb. The driving mode damping can be expressed as:

$$c_x = c_{xm} + c_{xcc} + c_{xcs} = \eta_p p\left(S_d + S_c + 2n_{dds}h_{damp}l_{damp} + 14n_{dds}\frac{b_{damp}h_{damp}^3}{a_{damp}^2}\right) \tag{6.10}$$

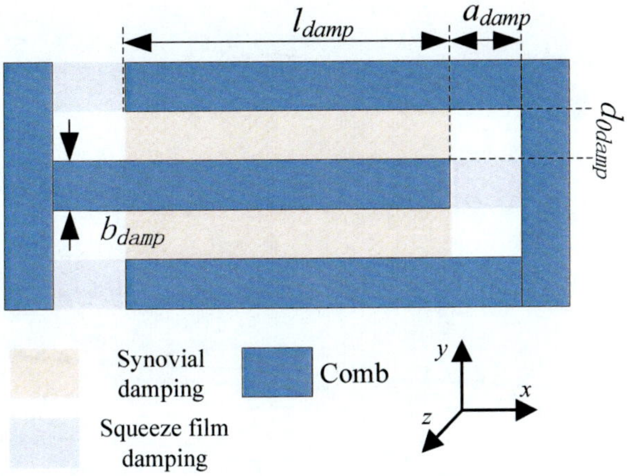

Fig. 6.2 Sliding film comb damping model

In the formula, c_{xm}, c_{xcc} and c_{xcs} are respectively the slide damping of the driving frame and the Coriolis mass with the glass substrate, the slide damping of the comb and the squeeze film damping of the comb; η_p is the air viscosity coefficient; p is the ambient air pressure; $n_{dds} = n_d + n_{ds}$ is the total number of driving mode comb teeth. The air in the package can be regarded as an ideal gas model. According to the Satherland formula and the Karabrone equation:

$$\frac{\eta_p}{\eta_{p0}} = \left(\frac{T}{T_0}\right)^{3/2} \frac{T_0 + B_0}{T + B_0} \tag{6.11}$$

$$V_{air}\, p = n_{air} R T \tag{6.12}$$

In the formula, η_p and $\eta_{p0} = 3.7 \times 10^{-4}$ kg / m^2 s·Torr are air viscosity coefficients at T and $T_0 = 300$ K; $B_0 = 120$ K is the Saserland constant; $R = 8.314$ J/mol K is the molar gas constant; v_{air} and n_{air} are the packing volume and the mole number of air. Substituting Eqs. (6.11) and (6.12) into Eq. (6.10), we can obtain:

$$c_x = \eta_{p0}\left(\frac{T}{T_0}\right)^{3/2} \frac{T_0 + B_0}{T + B_0} \frac{n_{air} R T}{V_{air}} \left(S_d + S_c + 2 n_{dds} h_{damp} l_{damp} + 14 n_{dds} \frac{b_{damp} h_{damp}^3}{a_{damp}^2} \right) \tag{6.13}$$

Substituting the corresponding parameters in Table 2.1, the proportion of each damping term in the total air damping can be obtained, as shown in Fig. 6.3. Although the comb structure adopts sliding film, the small range of pressure film damping between the top of the comb and the bottom of the opposite comb becomes the main part of air damping, accounting for about 70%; the film damping between Coriolis mass and driving frame and glass substrate accounted for 23 and 2% respectively. The parallel sliding film damping between the comb teeth accounts for 6%. Therefore, the tip or arc-top comb can be used in the design of gyro comb to effectively

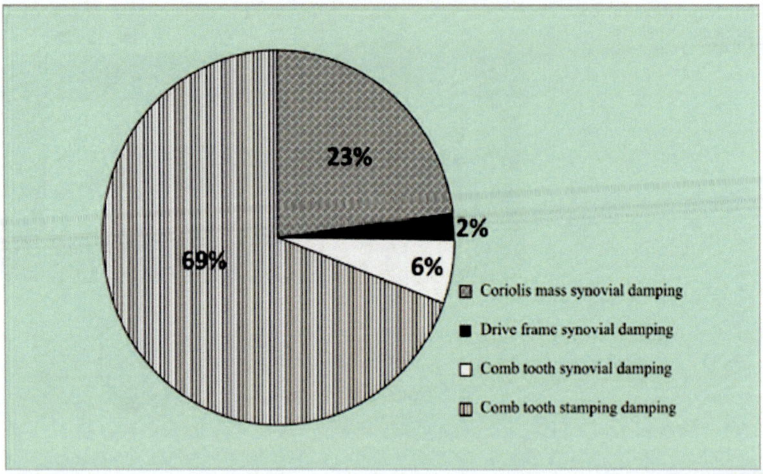

Fig. 6.3 Composition of damping components of driving mode

reduce the modal damping. In addition, the damping of the driving mode can be further expressed as:

$$c_x = A_{cx} \frac{n_{air}}{V_{air}} \frac{T^{5/2}}{T + 120} = \frac{\omega_x m_x}{Q_x} \tag{6.14}$$

where $A_{cx} = 8.68 \times 10^{-9}$ m s J / mol $K^{1.5}$ is the conversion coefficient of damping. Although the change of temperature will cause the change of silicon structure size, this change only accounts for 0.5‰ of A_{cx}, so the influence of structure size change on system damping can be ignored. It is not difficult to see from the above formula that the higher the temperature is, the greater the damping is and the lower the quality factor is when the packaging volume and the molar number of air are constant. Temperature and volume remain unchanged, the thinner the air (the smaller the mole number), the greater the quality factor.

Since the temperature has a great influence on the mechanical parameters of the silicon-based MEMS gyroscope structure, four packaged dual-mass linear vibration silicon-based MEMS gyroscope head samples (A, B, C, D) are selected to carry out the full temperature change experiment of the driving mode (−40 to 60 °C). The resonant frequency and quality factor of the driving mode are measured every 10 °C. Based on the above models of temperature, resonant frequency and quality factor, the mechanical parameters of driving modes at different temperatures are quantified. The test values and model quantification values are given in the chart. The temperature curves of the four samples are shown in Figs. 6.4, 6.5, 6.6 and 6.7.

It can be seen from the full temperature curves of the above four header samples that the relative variation of the resonant frequency of the driving mode is about 2‰ in the full temperature range, while the relative variation of the quality factor is about

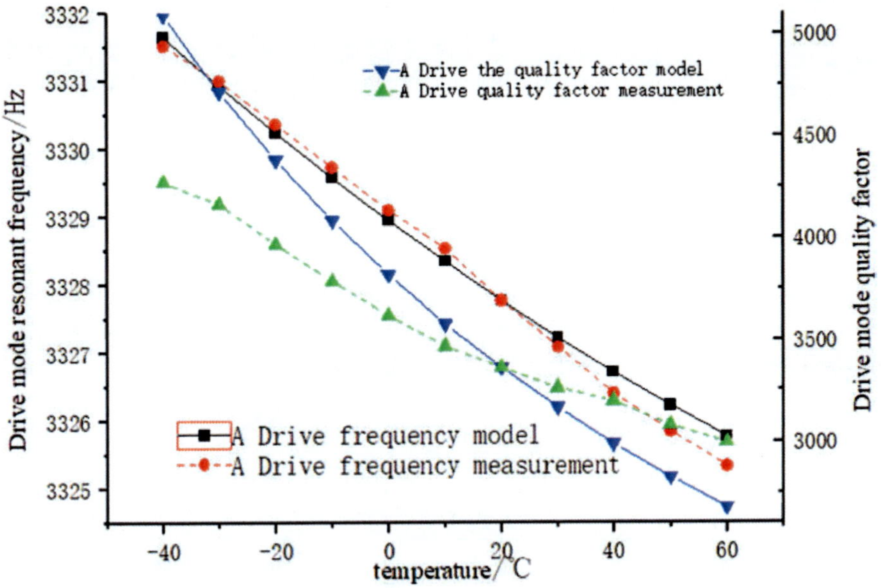

Fig. 6.4 Full temperature driving mode curve of head A

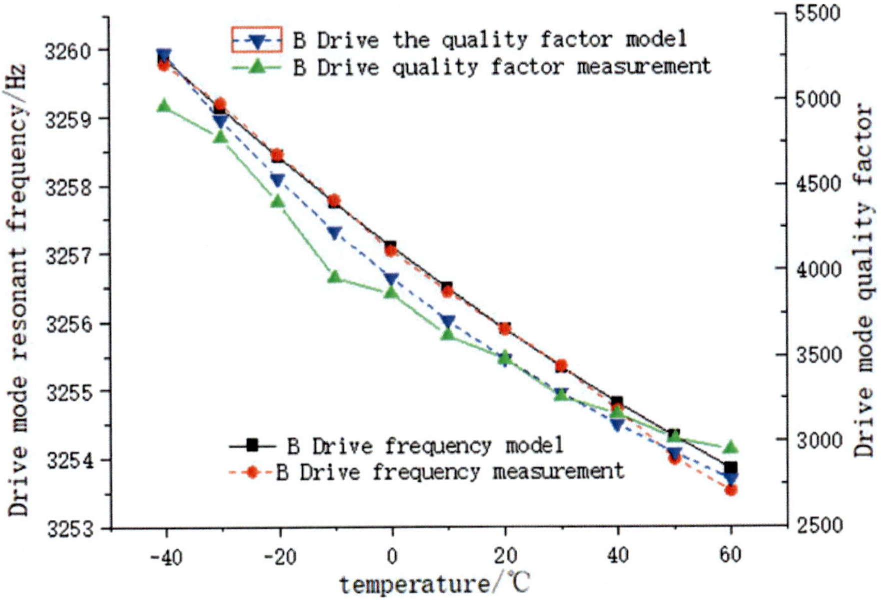

Fig. 6.5 Driving modal curve of meter head B at full temperature

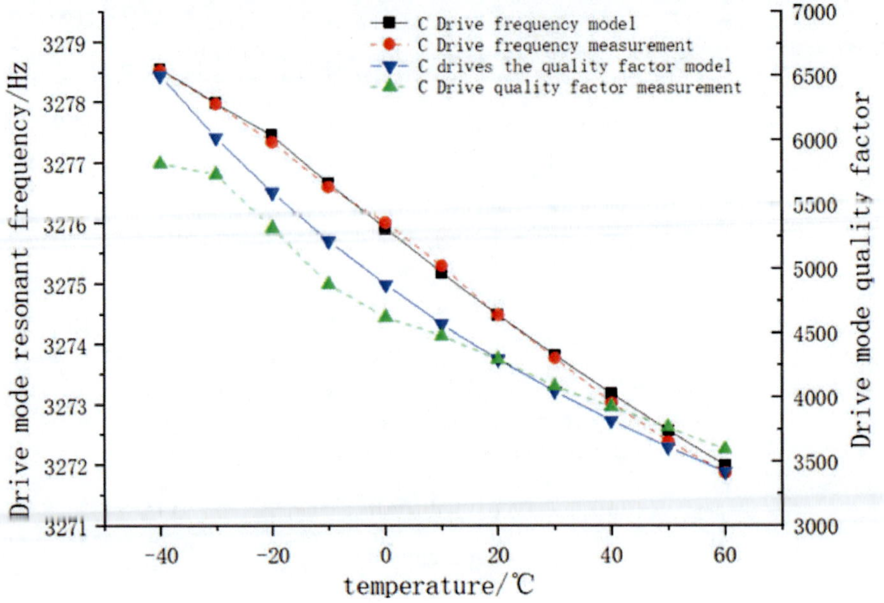

Fig. 6.6 Table Head C full-temperature driving modal curve

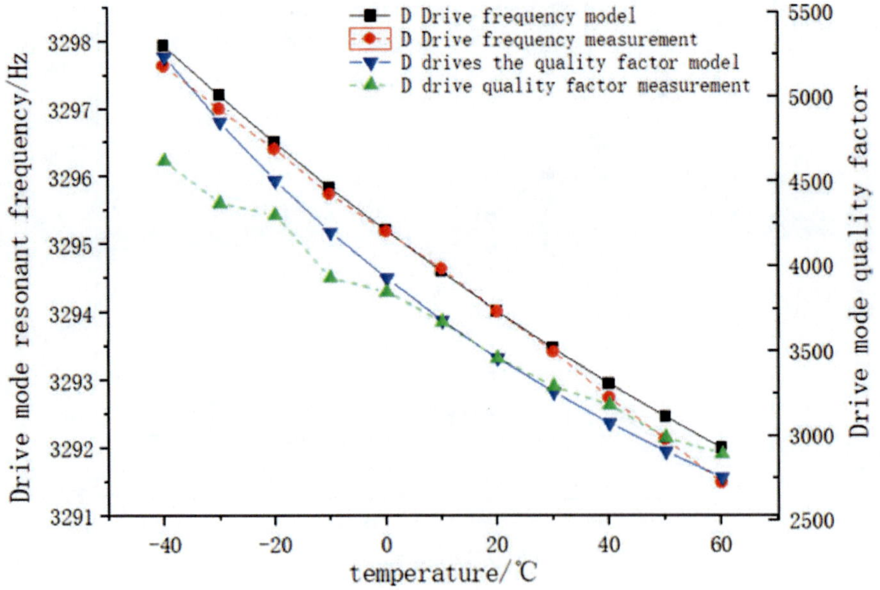

Fig. 6.7 Full temperature driving mode curve of head D

30–40%. The resonant frequency model given in this chapter is basically consistent with the measured data curve. At the same time, the model curve of quality factor can also roughly reflect the change trend of real quality factor. The model value of the quality factor is obtained as follows: Because the ratio of the amount of air in the package to the volume of space (n_{air}/V_{air} in Formula (6.14)) is difficult to measure, so the measured value of the quality factor at 20 °C is used as the benchmark to calculate the corresponding n_{air}/V_{air}. Since encapsulation can be regarded as a closed space, and n_{air} and V_{air} can be basically regarded as unchanged during the experiment, n_{air}/V_{air} can be calculated as a constant.

6.2.3 *Effect of Temperature on Zero Bias*

Zero bias is the direct output of the silicon-based MEMS gyroscope without angular velocity input signal. Without changing the position and static state of the gyroscope, only changing the ambient temperature can obtain the curve of the influence of temperature on zero bias. The following Figure is the curve of a dual-mass linear vibration silicon-based MEMS gyroscope changing with temperature, where the red curve is the temperature signal collected by the temperature sensor in the gyroscope in real time. It can be seen that the gyro bias changes from 5 to 7.5°/s from −30 to 50 °C (Fig. 6.8).

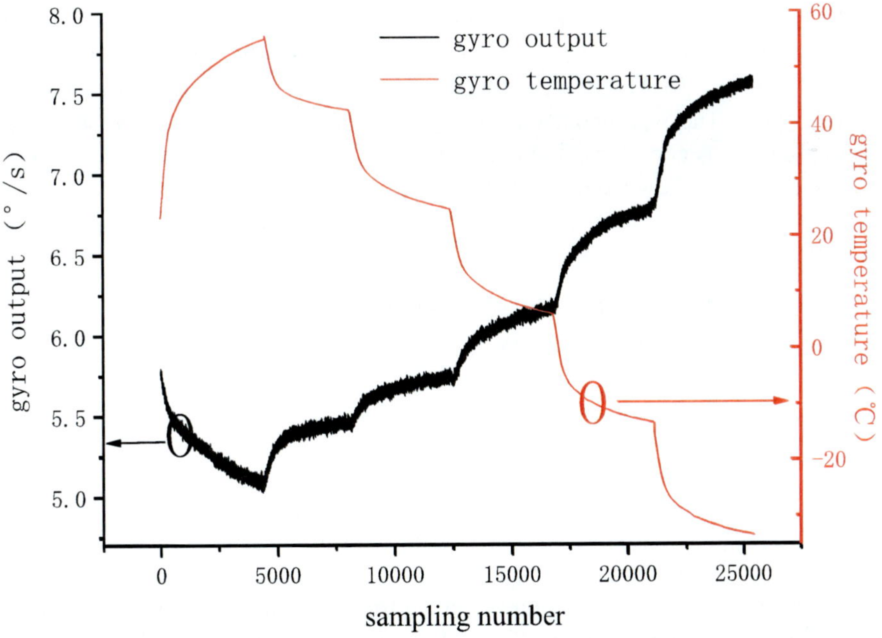

Fig. 6.8 Dual-mass linear vibration silicon-based MEMS gyroscope zero Bias temperature curve

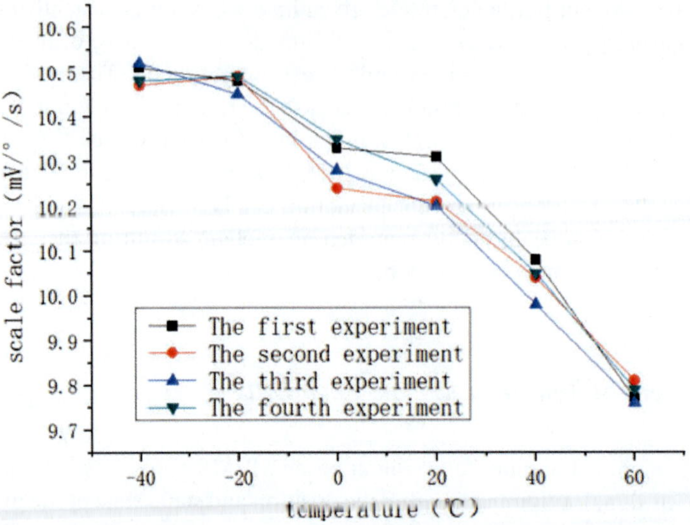

Fig. 6.9 Scale factor and temperature curve of dual-mass linear vibration silicon-based MEMS gyroscope

6.2.4 Temperature Influence on Scale Factor

The scale factor is another important parameter that reflects the sensitive input angular rate of gyroscope. The temperature coefficient of the scale factor of a dual-mass linear vibration silicon-based MEMS gyroscope is 693 ppm/°C in the range of −40 to 60 °C (as shown in Fig. 6.9). Three repeatability experiments are carried out under the same test conditions. The experimental results also prove that the scale factor parameter of the gyroscope has good temperature repeatability, and also reflect that the influence of temperature on the scale factor cannot be ignored.

6.3 Silicon Based MEMS Gyroscope Temperature Controlling Technology

6.3.1 Chip-Level Temperature Controlling Technology [157, 158]

Usually, the silicon-based MEMS gyroscope sensitive structure is encapsulated in a cavity, on the one hand to protect the sensitive structure, on the other hand can also provide a vacuum working environment for the sensitive structure, reduce the air damping coefficient, improve the quality factor and mechanical sensitivity. The

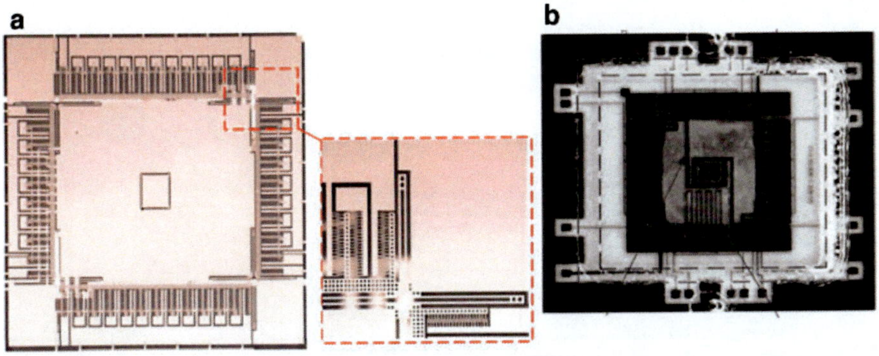

Fig. 6.10 Diagram of distribution of the silicon-based MEMS gyroscope structural layer (**a**) and heating layer (**b**)

chip-level temperature control technology is to use temperature sensors and resistance heaters in this small vacuum chamber to monitor the temperature, stabilize the temperature in the vacuum chamber, and reduce the influence of external environmental temperature changes on the silicon-based MEMS gyroscope sensitive structure. This technology not only shortens the lag time of the temperature model, reduces the power consumption and improves the temperature control accuracy, but also can be integrated with the silicon-based MEMS gyroscope structure to facilitate mass production. With silicon as the base, the glass insulation layer with poor thermal conductivity is bonded on the base. The micro-heating wire RL and micro-temperature sensor (thermal resistance) RS are processed with platinum-titanium alloy material and distributed on the insulation layer. The gyroscope sensitive structure is located at the top (Fig. 6.10), and the whole structure is encapsulated in the ceramic vacuum chamber.

In order to verify the heating performance and heat distribution of the designed structure, Ansys software is used to simulate the temperature distribution of the heating layer and the silicon-based MEMS gyroscope structure layer (Fig. 6.11).

It can be seen that the heating wire can provide the necessary temperature for the silicon-based MEMS gyroscope structure. The temperature difference between the center temperature and the surrounding temperature of the gyro structure is within 1 °C, and the temperature distribution is uniform. The whole structure of chip temperature control gyroscope is processed by SOG process. Firstly, four 80um anchor points are formed on 200um thick silicon substrate, and metal is deposited on glass substrate. Secondly, platinum titanium alloy microwires and temperature sensors were deposited on 300 um thick glass substrate. Then, the glass substrate is bonded with the silicon substrate, and then the silicon substrate is thinned and polished. Finally, the release structure of silicon substrate was treated with DRIE technology. The whole process is shown in Fig. 6.12.

When the temperature control system works, the micro heating wire RL generates heat under the action of external driving voltage, and the heat diffuses in the packaging cavity by radiation. Due to the poor thermal conductivity of the glass layer at the

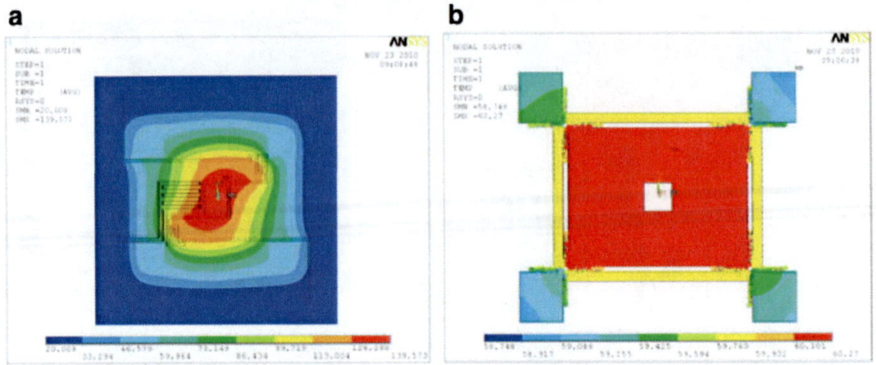

Fig. 6.11 Simulation temperature distribution heating layer (**a**) and silicon-based MEMS gyroscope structure layer (**b**)

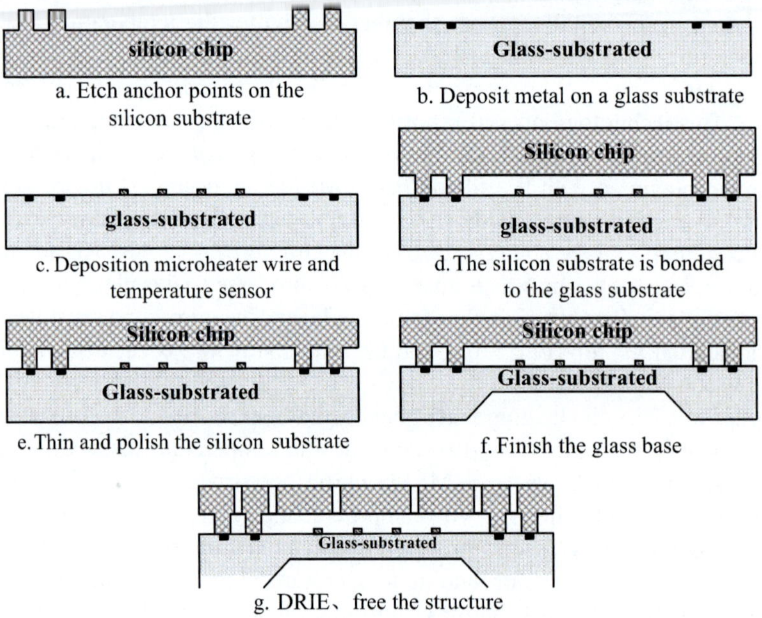

Fig. 6.12 Chip-level temperature control silicon-based MEMS gyroscope processing technology

bottom, most of the heat is absorbed by the gyro structure above the heating wire. The temperature sensor RS is a thermal resistor, which is sensitive to the temperature in the cavity and outputs to the interface of the gyro surface head, providing the temperature information in the cavity for the control circuit. The circuit of temperature control system includes thermistor signal extraction circuit, PID control circuit (the output set a diode to prevent negative voltage heating), voltage reference circuit, power drive circuit. The system block diagram is shown in Fig. 6.13.

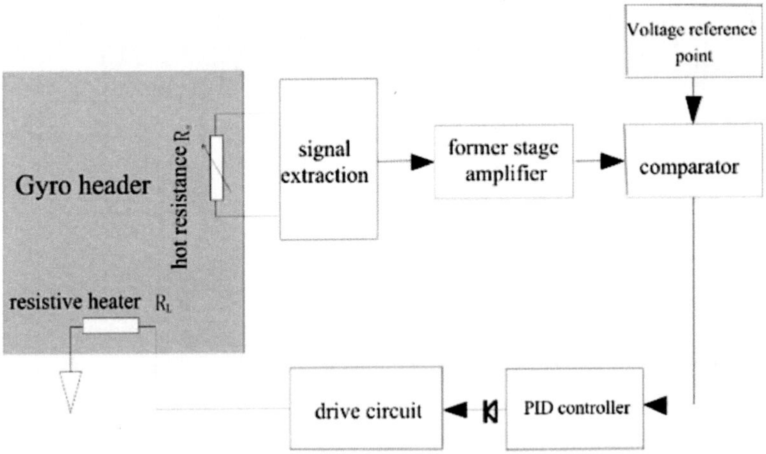

Fig. 6.13 Principle diagram of chip temperature control system

The whole loop consists of an automatic gain control system. The temperature point inside the chip is controlled by the voltage reference. When the temperature of the front panel is less than the set value, the absolute value of the voltage output by the front amplifier module is less than the voltage reference. The comparator output positive voltage makes the PID controller output positive, the diode is on, the drive circuit works, the heating resistance generates heat, and the temperature in the front panel increases. When the temperature in the header exceeds the set value, the amplification signal of the front stage is larger than the voltage reference. The output signal of the comparator is opposite. The output of the PID controller is negative, and the diode is cut off, resulting in zero signal at the input end of the driving circuit. In this case, the driving voltage of the heating wire is very small (about several mV), which cannot effectively generate heat. At the same time, with the outward diffusion of heat in the header, the temperature in the header decreases. Under the action of this dynamic adjustment, the internal temperature of the header will eventually be stabilized near the set value, so as to achieve the purpose of temperature control. It should be noted that since the heating wire RL can only control the temperature by heating, the working temperature of the gyroscope head should be set above the upper limit (60 °C) of the external environment test temperature to ensure that the system is always in the heating working state within the range of external temperature. In order to minimize the heating power consumption, the working temperature of the gyro head is set at 65 °C. The step response and system modeling and simulation are also based on this temperature.

In the aspect of controller design, the relationship model between the heating voltage at both ends of RL and the temperature information detected by RS is established. Because the temperature control system is a first-order inertial plus pure delay system, the transfer function is:

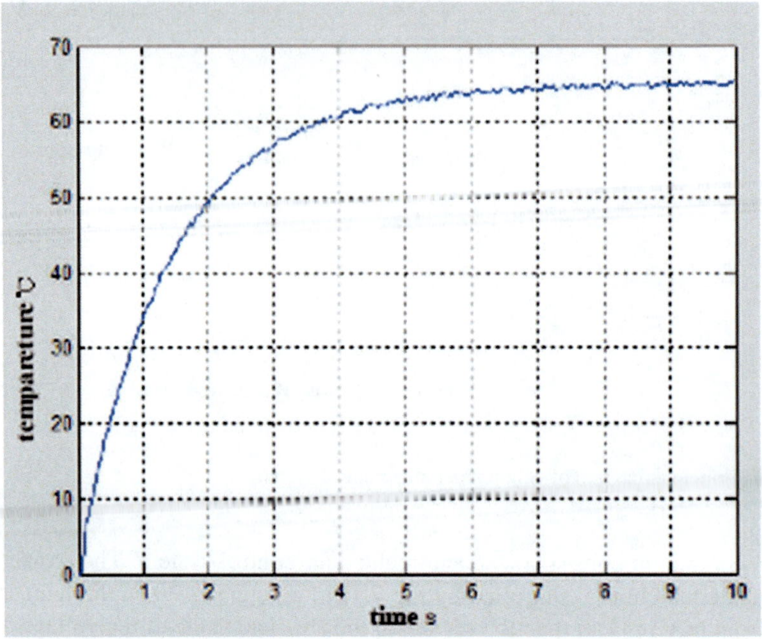

Fig. 6.14 Step response diagram of temperature in the head of chip temperature control silicon-based MEMS gyroscope

$$G_{\mathrm{ctc}}(s) = \frac{K_{\mathrm{ctc}} e^{-L_{\mathrm{ctc}} s}}{T_{\mathrm{ctc}} s + 1} \tag{6.15}$$

Among them, K_{ctc} is the amplification factor, T_{ctc} is the inertia time constant, L_{ctc} is the delay time. Step response method is used to determine the above parameters, the process is as follows: Under the condition of 0 °C, step voltage is added at both ends of R_L, the resistance measured by R_S is converted into temperature signal, and the step response diagram is drawn (Fig. 6.14).

It can be seen from the figure that the system can reach a stable state within 10 s, and the system delay is small, which is conducive to the design of temperature control system. Based on the feature area method, the curve area feature is extracted from the above graph, and the parameters K_{ctc}, $T_{ctc,}$ Lctc are obtained according to the Formula (6.16):

$$\begin{cases} K_{\mathrm{ctc}} = \dfrac{y_{\mathrm{ctc}}(\infty)}{u_{0\mathrm{ctc}}} \\ T_{\mathrm{ctc}} = \dfrac{e T_{0\mathrm{ctc}}}{y_{\mathrm{ctc}}(\infty)} \displaystyle\sum_{i=0}^{m_{r\mathrm{ctc}}} y_{\mathrm{ctc}}(t) \\ L_{\mathrm{ctc}} = \dfrac{T_{0\mathrm{ctc}}}{y_{\mathrm{ctc}}(\infty)} \displaystyle\sum_{i=0}^{m_{\mathrm{ctc}}} [y_{\mathrm{ctc}}(\infty) - y_{\mathrm{ctc}}(t)] - T_{\mathrm{ctc}} \end{cases} \tag{6.16}$$

Table 6.2 Ziegler-Nichols empirical parameter formula

Controller type	K_{Pctc}	T_{Ictc}	T_{Dctc}
P	$1/\alpha_{ctc}$		
PI	$0.6/\alpha_{ctc}$	$3L_{ctc}$	
PID	$1.2/\alpha_{ctc}$	$2L_{ctc}$	$0.5L_{ctc}$

Table 6.3 Simulation parameters

Parameter	Numerical value	Parameter	Numerical value
K_{ctc}	13	C/°C	65
T_{ctc}	3.2417	K_{Pctc}	0.085527
L_{ctc}/s	0.34987	T_{Ictc}/s	0.69974
K_{0ctc}	1	T_{Dctc}/s	0.174935

In the formula, y_{ctc} (∞) is the steady-state value of the system 65 °C, u_0 is the step voltage input value 5 V, e is the bottom of the natural logarithm, T_{0ctc} is the sampling period 0.02 s, m_{rctc} is the time required for the system to rise to the steady-state value, y_{ctc} (t) is the temperature value at t time, m_{ctc} is the total number of sampling points. PID parameters of temperature control system is determined by Ziegler-Nichols empirical parameter method, such as Table 6.2.

Among them, $\alpha = (\text{Kctc}\ L_{ctc}/T_{ctc})$ is a constant related to the controlled object model.

According to the principle of temperature control system, the system model is established in Matlab and the parameters in Table 6.3 are introduced. The simulation results are shown in Fig. 6.15. The system can stabilize the temperature at 65 °C in a short time.

In order to reflect the real temperature inside the gyroscope head and facilitate the establishment of temperature model, RS is calibrated in this book: from −20 to

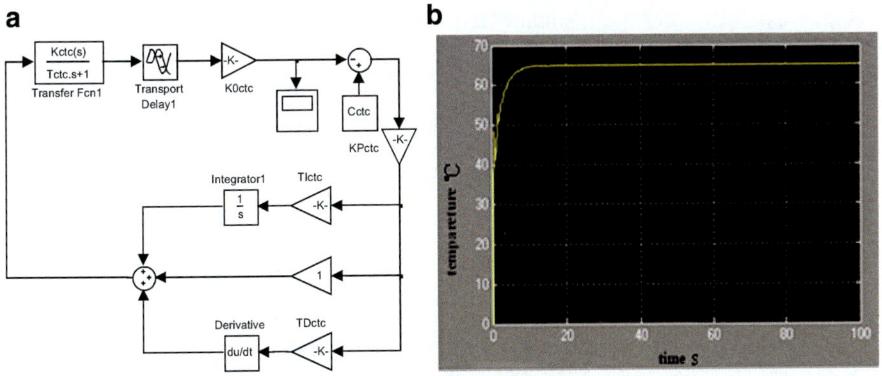

Fig. 6.15 Simulation model of chip-level temperature control system (**a**) and simulation results (**b**)

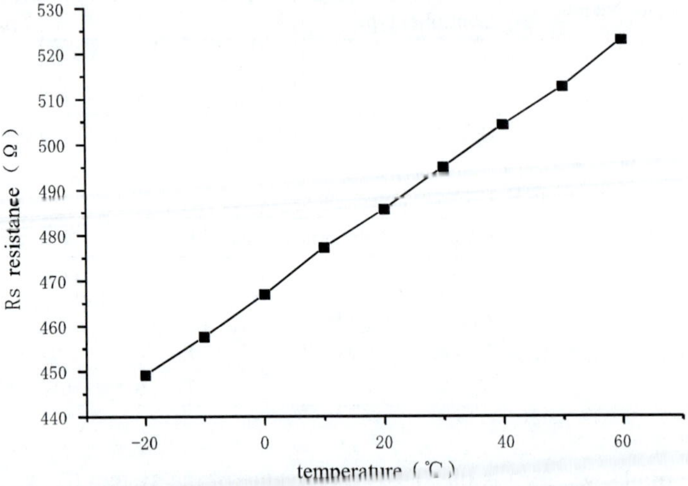

Fig. 6.16 Silicon-based MEMS gyroscope temperature characteristic curve of temperature sensor R_S

60 °C, the RS resistance is measured every 10 °C, and the curve of its change with temperature is shown in Fig. 6.16.

It can be seen from the Figure that thermistor RS has a good linearity in the whole temperature range, and the equation of the relationship between thermistor value and temperature t can be obtained by least square fitting (Eq. 6.17):

$$R_S = 466.9 \times (1 + 0.002 \times t) \tag{6.17}$$

After the temperature control system is added, the thermal imaging of the surface of the head is taken in the atmospheric state, as shown in Fig. 6.17. Subsequently, the gyroscope was tested from −20 to 60 °C before and after the temperature control system was added, and the resonant frequency f_d of heating resistance and driving mode was recorded every 20 °C (Fig. 6.18).

It can be seen from the above diagram that the temperature change inside the silicon-based MEMS gyroscope head has been significantly improved after the temperature control system works. The variation of thermistor in the whole temperature range decreases from nearly 77 Ω before temperature control to less than 5 Ω after temperature control. The numerical values in Fig. 6.18 are introduced into the Formula (6.17), and the temperature range in the header package is obtained: from −19.987 to 58.466 °C, and the change is 78 °C. After the temperature control, the internal temperature of the gyro head changed from 66.510 to 61.561 °C, only 4.95 °C, which greatly stabilized the internal temperature of the head. Then the driving mode resonance frequency f_d of the gyro head changed from 3.76 Hz before the temperature control to 0.48 Hz after the temperature control.

Silicon material is very sensitive to temperature change, physical parameters such as surface structure size and Young modulus will change with temperature, which

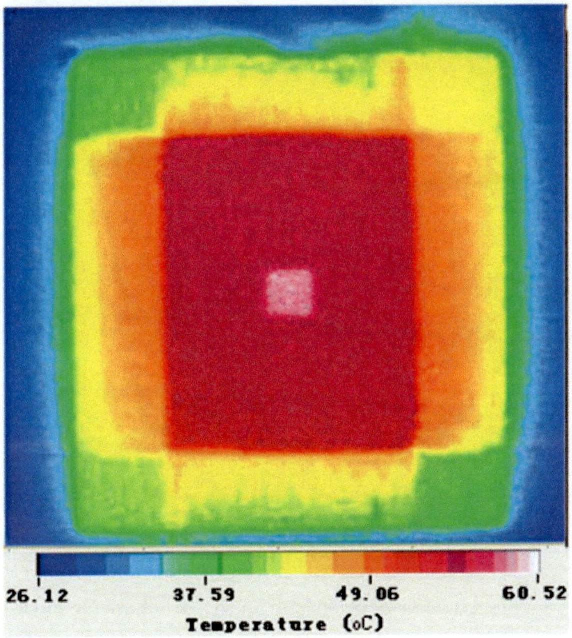

Fig. 6.17 Thermal imaging photos of the head of the temperature control system

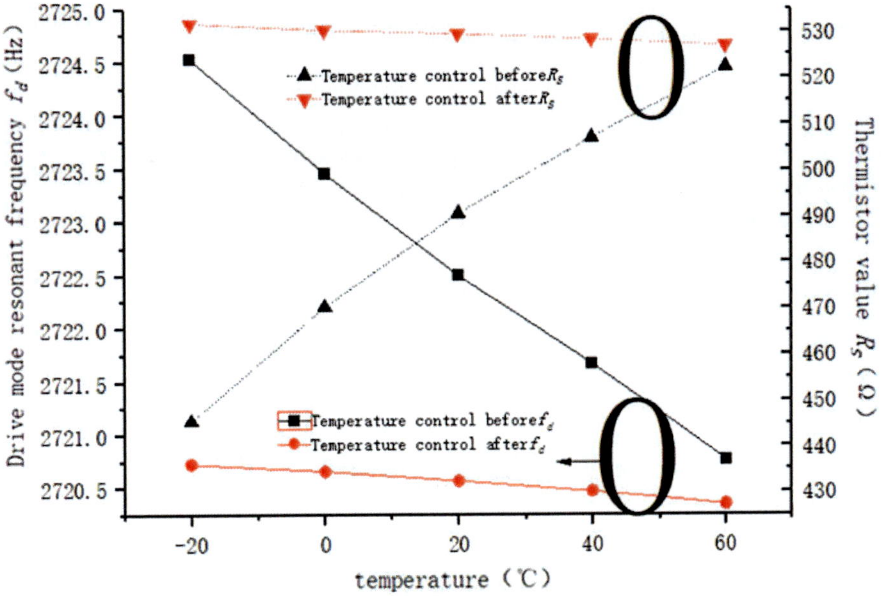

Fig. 6.18 Comparison of R_S and f_d total temperature effects before and after temperature control

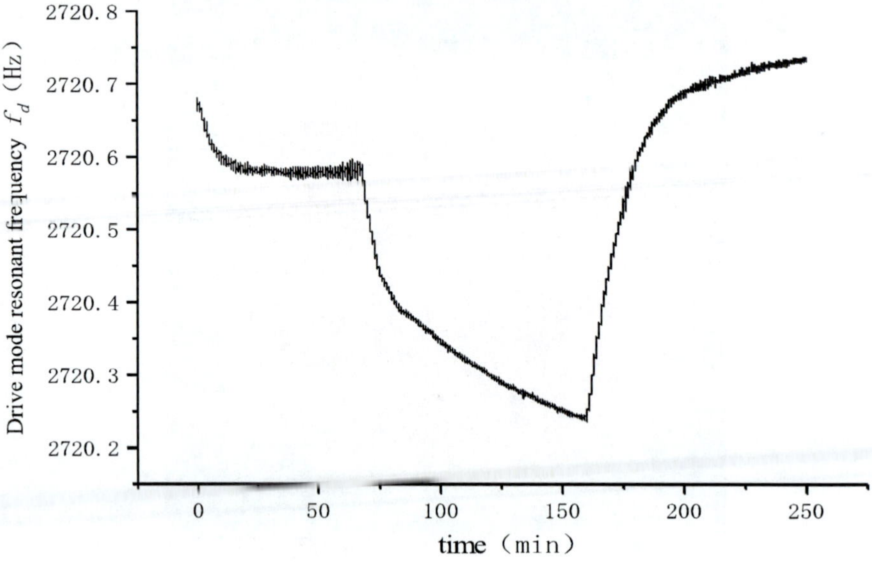

Fig. 6.19 Full temperature test curve of resonant frequency f_d of driving mode after temperature control

has an impact on the stiffness coefficient of the supporting beam, resulting in the resonant frequency f_d of the driving mode varies greatly in the whole temperature range. After adding the temperature control system, the f_d tends to be stable in the whole temperature range because the temperature around the silicon structure changes little. In order to verify the stability of the temperature control system in the normal temperature and full temperature range of the chip-level temperature control silicon-based MEMS gyroscope, the gyroscope is started at normal temperature for one hour under the condition of complete cooling of the gyroscope, and the temperature is increased to 60 °C and stabilized, and then the temperature is decreased to -20 °C. The f_d of the whole process is recorded as Fig. 6.19. Since the test circuit generates heat after it starts to work, f_d tends to be stable as the temperature in the silicon-based MEMS gyroscope gradually reaches equilibrium.

6.3.2 Sensor—Level Temperature Controlling Technology [162–164]

The chip-level temperature control technology introduced in the above section is directly aimed at the temperature control of the silicon-based MEMS gyroscope structure inside the gyro structure packaging. Because the structure itself does not produce temperature in the working process, and the structure area is small, the realization of the control system is easier. The whole-scale temperature control technology introduced in this section is carried out inside the outermost shell of the silicon-based

MEMS gyroscope. The temperature control objects are the gyroscope measurement and control system and the outer structure of the packaging. Since the measurement and control system will generate temperature after power, and the temperature distribution is not uniform, these practical problems need to be considered in the whole-scale temperature control. The temperature inside the whole gyroscope shell is detected by a mature temperature sensor (such as DS18B20). The temperature controller is realized by a semiconductor refrigeration chip, which is installed on the whole gyroscope shell, as shown in Fig. 6.20. Similarly, the classical and easily realized PID control method can be used to control the temperature control system. Firstly, the temperature model of the system is identified, and the system model is determined as shown in Eq. (6.18). According to the simulation, the PID parameters are determined as $P_{Stc} = 1$, $I_{Stc} = 0.03$, and $D_{Stc} = 0.1$.

$$G_{Stc}=\frac{42}{2735s+1}e^{-1350s} \tag{6.18}$$

When the system starts to work, the temperature reading instruction is first issued by the single chip microcomputer, and the current temperature value T_{1Stc} of the silicon-based MEMS gyroscope is measured by the digital temperature sensor DS18B20. Then, T1Stc is compared with the set value T_{Stc}, and the difference is sent to the PID controller. The PID controller outputs a certain number of control values after processing, which is converted into analog voltage by D/A. The voltage signal is then driven by a large current drive circuit to improve the current drive ability and loaded on the semiconductor cooler to heat or cool the temperature control object. In order to speed up the control, the program design uses integral separation PID control: setting threshold Δt, when the temperature difference is greater than Δt, using PD control; when the temperature difference is less than or equal to Δt, it enters the PID control link. The software flow of temperature control main program is shown in Fig. 6.21. According to the above design, the system is built and

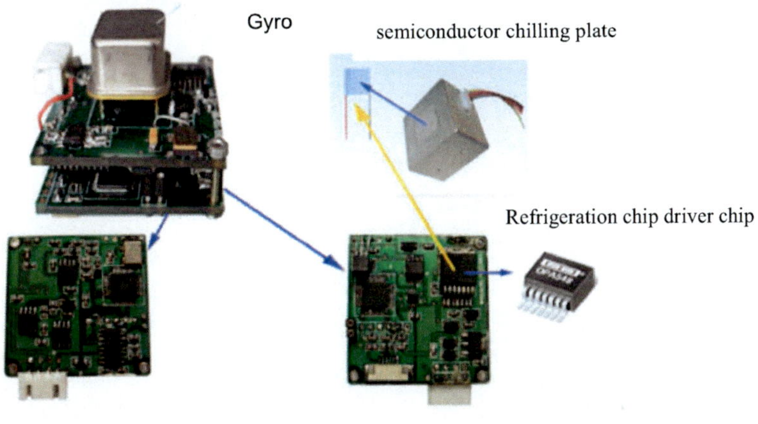

Fig. 6.20 Temperature control system photos

tested. The curves of ambient temperature and temperature inside the silicon-based MEMS gyroscope table are shown in Fig. 6.22, and the zero output of the gyroscope is shown in Fig. 6.23. It can be seen that when the external temperature is from −30 °C to above 40 °C, the internal temperature of the gyroscope is always stable at about 55 °C, and the bias of the gyroscope is also relatively stable. The temperature control technology of the whole table level can ensure that the gyroscope is in a relatively constant temperature environment, which greatly improves the ability of the silicon-based MEMS gyroscope to resist temperature changes.

Fig. 6.21 Working flow chart of the silicon-based MEMS gyroscope temperature control system

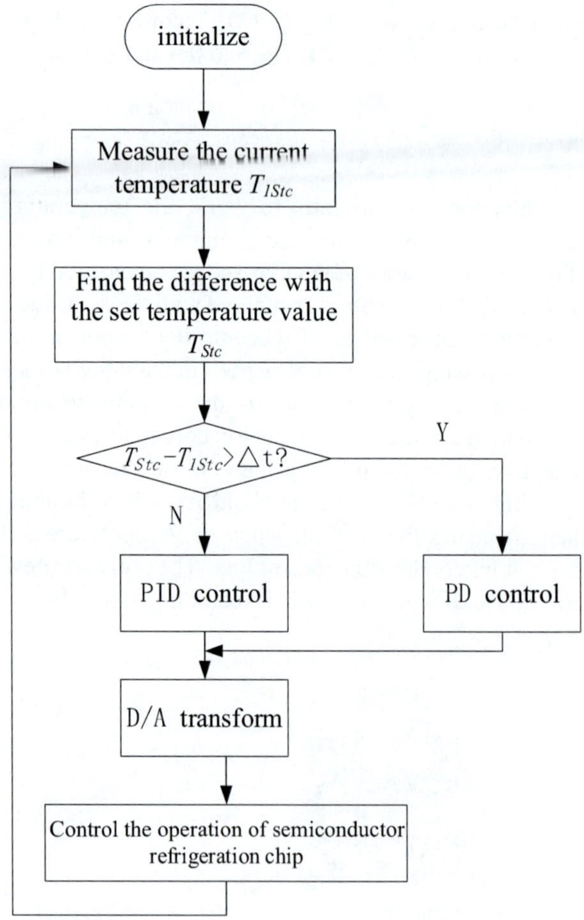

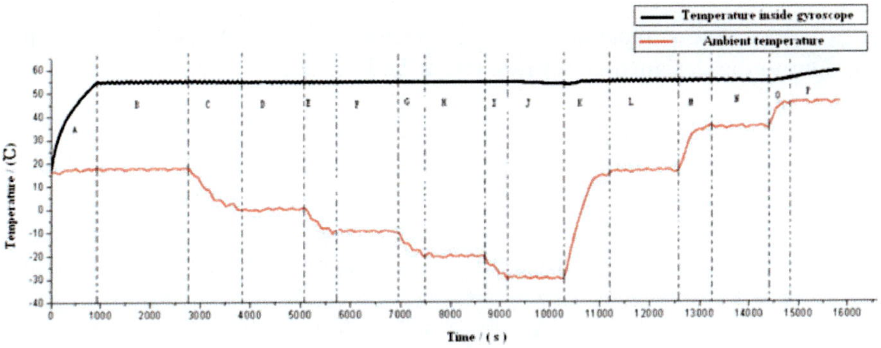

Fig. 6.22 Internal and external temperature test curves of the silicon-based MEMS gyroscope during operation of temperature control system

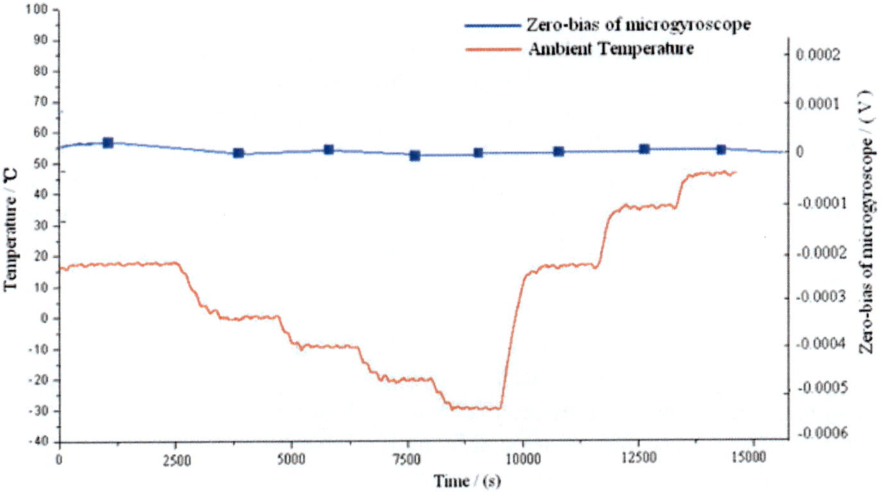

Fig. 6.23 Gyro output and external temperature test curve of the silicon-based MEMS gyroscope temperature control system

6.4 Silicon Based MEMS Gyroscope Temperature Compensation Technology

The temperature compensation technology of silicon-based MEMS gyroscope does not require additional heating or cooling modules to control the internal temperature of the gyroscope, which avoids the temperature control system and reduces the complexity of the whole gyroscope system. On the other hand, it also reduces the power consumption of the system, which is suitable for low power applications. The temperature compensation technology uses the temperature sensor or other information (such as modal resonant frequency) to sensitive the internal temperature of the gyroscope, and compensates the scale factor and zero bias of the gyroscope according to the designed compensation method, so as to reduce the variation of the scale factor and zero bias in the whole temperature range. However, compared with temperature control technology, temperature compensation technology also has disadvantages. Firstly, this technology can only be used for gyroscope individuals with good repeatability and certain temperature variation trend. Secondly, this technology is not universal, and it must be matched by accurate compensation parameters according to different gyroscope individuals to finally form the compensation purpose.

6.4.1 Temperature Compensation Technology Based on Temperature Sensor

The premise of temperature compensation is to obtain the temperature inside the silicon-based MEMS gyroscope. The most direct method is to add a temperature sensor to the gyro measurement and control circuit, and compensate the relevant parameters according to the temperature value on several ‘key points’ to finally achieve the purpose of affecting the key parameters of the silicon-based MEMS gyroscope (scale factor and zero bias). This section takes HAYASHI DENKO platinum resistance temperature measuring element as an example to introduce temperature compensation technology.

The thin-film platinum resistance is that platinum is sputtered on the ceramic substrate by the thin-film technology of vacuum deposition, and the thickness of the film is within 2 μm. The Ni (or Pd) lead is fixed by glass sintering material, and the thin-film element is prepared by laser resistance adjustment.

The platinum resistance of 1000 Ω is used in this book. Its resistance varies linearly with temperature. The formula can be expressed as follows:

$$R_T = R_0(1 + 0.0039t) \tag{6.19}$$

where R_T is the resistance value of temperature sensor; $r_0 = 1000\,\Omega$. The temperature characteristic curve of platinum resistance can be obtained from the Formula (6.19) as shown in Fig. 6.24.

Platinum resistance has two main functions:

1. Since the silicon-based MEMS gyroscope and test circuit are packaged into a shell and then put into the temperature control box for experiment, the temperature in the temperature control box cannot accurately reflect the actual temperature inside the silicon-based MEMS gyroscope. Therefore, the platinum resistance can be used as a temperature sensor to observe the change of resistance value through a multimeter, so as to obtain the temperature value according to Formula (6.19) and monitor the actual temperature inside the silicon-based MEMS gyroscope in real time.

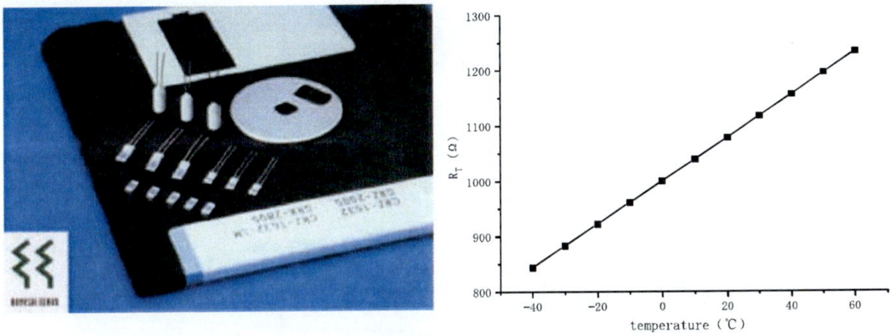

Fig. 6.24 Photograph of platinum resistance temperature sensor and temperature-resistance curve

2. The compensation methods studied in this paper are to find the key points that can affect the working state of the silicon-based MEMS gyroscope, and then design the compensation circuit according to the temperature coefficient of platinum resistance. Therefore, platinum resistance is the key component used in the compensation method.

In the compensation process, first of all, the silicon-based MEMS gyroscope parameters (scale factor and zero bias) need to be compensated for repeated temperature characteristics experiments to investigate the temperature repeatability and temperature coefficient; secondly, the whole measurement and control circuit of gyroscope is analyzed, and the key points that can affect the working state of gyroscope are found. Then, the compensation circuit is designed according to the temperature coefficient of platinum resistance. The circuit is simulated according to the nature of the compensation link, and finally the temperature test is carried out after the actual circuit is realized. It is worth noting that the simulation results can only judge the trend after compensation, and can not determine the specific value, because the introduction of compensation will have an impact on the whole system, not just the compensation itself in the module. After the temperature compensation of scale factor, the compensation method of zero offset output can be considered. The zero-bias output balances the full-temperature output by superposition with the temperature compensation link, and the opposite trend offsets the influence of temperature by adding thermistor in the measurement and control circuit.

The process of temperature compensation scheme is shown in Fig. 6.25. Since the zero bias value will be changed in the process of scale factor compensation, the temperature compensation of scale factor should be carried out first. Firstly, repeated temperature experiments are carried out to find out the change trend and value of scale factor and zero deviation affected by temperature. If the repeatability of scale factor and zero deviation temperature is poor, the next temperature compensation cannot be carried out. Under the premise of good repeatability of scale factor and zero-bias temperature, the compensation scheme is designed according to the characteristics of platinum thermal resistors for several 'key points' that can affect the change of scale factor. After the temperature repeatability test of different schemes, the optimal

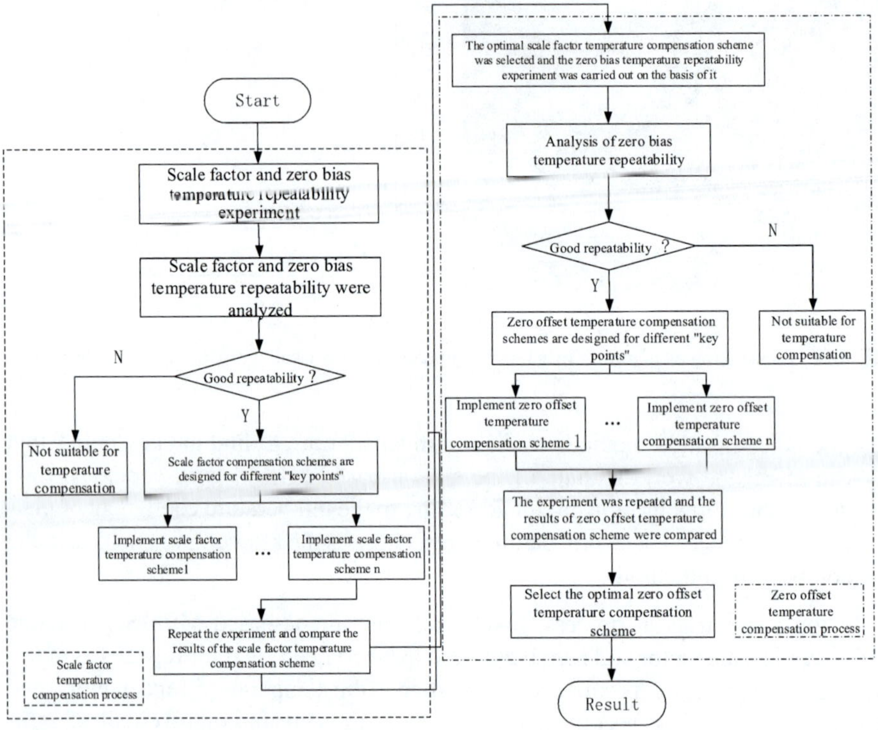

Fig. 6.25 'Key points' distribution of temperature compensation scheme based on temperature sensor

scheme is selected for the next zero-bias temperature compensation. Similarly, the temperature repeatability experiment is carried out on the zero bias. The 'key point' is selected for the gyroscope with good repeatability. The compensation methods are designed for different 'key points'. Finally, the best compensation scheme is selected for the repetition experiment of different compensation methods.

The repeated test curves of the scale factor and zero bias temperature of a dual-mass linear vibration silicon-based MEMS gyroscope are shown in Fig. 6.26. The temperature experiment process is as follows: the temperature characteristics of the gyroscope are tested: the gyroscope is powered up at room temperature, heated to 60 °C, and then cooled by 20 °C until −40 °C each time. The temperature is kept at each temperature point for 40 min to ensure that the internal and external temperatures of the gyroscope shell are the same. The zero bias is collected and output for 60 min, and the scale factor is measured. The temperature coefficients of the scale factor and zero bias in the whole temperature range are 393 ppm/°C and 75°/h/°C (the scale factor is 9.23 mV/°/s). The full-temperature experiments of these two parameters were carried out for three times, and the results showed that the repeatability was good. The least square fitting was performed after taking the mean values of the three groups of data for subsequent compensation calculation. According to the

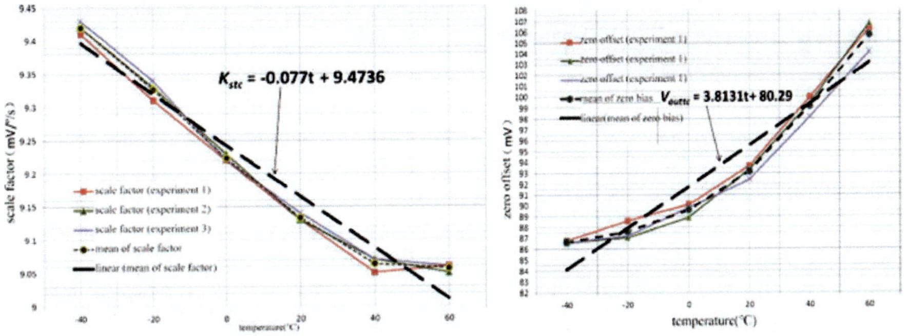

Fig. 6.26 Repeated test curves of scale factor and zero bias temperature for a dual-mass linear vibration silicon-based MEMS gyroscope

compensation process, the 'key points' of scale factor compensation are selected as follows: According to the analysis Formulas (3.33) and (3.34) of the gyro drive circuit in the third chapter and the mechanical sensitivity expression (2.30) in the second chapter, it is determined that the voltage reference V_{ref} and the amplification link K_{PX} in the amplification phase shifter can affect the mechanical sensitivity of the gyro. At the same time, the K_{PY} in the detection circuit can also affect the scale factor by gain. The above three 'key points' achieve the purpose of compensation by adjusting the working point of the drive circuit, the gain of the drive circuit and the gain of the detection circuit.

This book takes the adjustment of K_{PY} as an example to introduce the design of temperature compensation circuit in detail. In order to ensure that the detection circuit has sufficient gain and high phase accuracy, the circuit diagram of the module is shown in Fig. 6.28, and the transfer function is:

$$K_{PY} = \frac{v_{otc}}{v_{itc}} = \frac{(R_{1tc} + R_{2tc} + R_{ttc})R_{3tc}}{R_{1tc}(R_{3tc} + R_{4tc})} \tag{6.20}$$

where R_{ttc} is RT in Eq. (6.19). Since the temperature coefficient of the scale factor shown in Fig. 6.27 is −0.077, K_{PY} should provide a positive temperature coefficient of 0.077 to minimize the temperature effect of the scale factor. According to the Formula (6.19), the circuit parameters in Table 6.4 can be used as compensation parameters. In the table, 'before compensation' refers to the resistance fixed value without temperature compensation.

After substituting the values in the table, the temperature characteristics of scale factor were simulated, and three repeatability experiments were carried out. The results are shown in Fig. 6.29. The simulation results are roughly consistent with the actual measurement results, and the repeatability is good. The temperature variation of scale factor is about 73 ppm/°C (K_S = 9.20 mV/o/s). At the same time, the temperature repeatability test of zero bias is carried out. The temperature coefficient of zero bias is 115.6°/h/°C and has a good linear trend, which is conducive to the next compensation.

Table 6.4 KPY module and FLPF1 module circuit parameters

Parameter	Resistance values (Ω)		Parameter	Resistance values (Ω)	
	Doppler compensation	After compensation		Doppler compensation	After compensation
R_{1tc}	10 k	1.5 k	R_{5tc}	∞	1.5 k
R_{2tc}	30 k	7.5 k	R_{6tc}	∞	20 k
R_{3tc}	100 k	50 k	R_{8tc}	∞	R_{7tc}
R_{4tc}	10 k	10 k	R_{ttc2}	1 k	R_{ttc}
R_{ttc}	1 k	R_{ttc}			

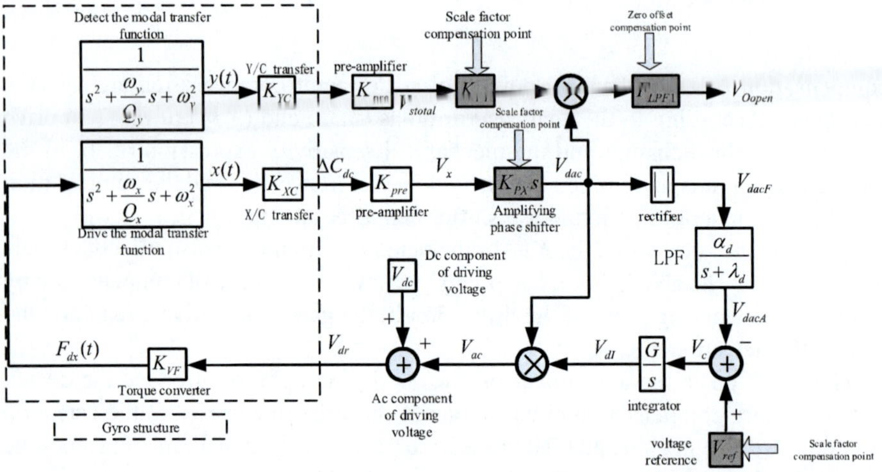

Fig. 6.27 'Key Points' distribution of temperature compensation scheme based on temperature sensor

Fig. 6.28 K_{PY} module circuit diagram

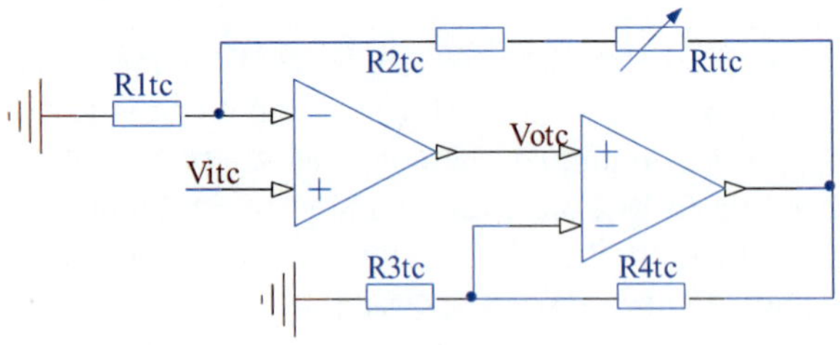

For the zero-bias temperature compensation, the low-pass filter F_{LPF1} of the output stage is selected, so that not only the original zero-bias temperature drift can be offset by voltage superposition, but also the influence on the whole system is minimal. The circuit of the module is expanded as shown in Fig. 6.30, in which the virtual frame is the compensation circuit, and the transmission function satisfies Formula (6.21). V_{rdctc}, R_{12tc} and R_{13tc} constitute a zero-bit compensation circuit for zero offset nulling.

$$\frac{V_{reftc} \times \left(R_{6tc} + R_{ttc2(60^\circ)}\right)}{R_{5tc} + R_{6tc} + R_{ttc2(60^\circ)}} - \frac{V_{reftc} \times \left(R_{6tc} + R_{ttc2(-40^\circ)}\right)}{R_{5tc} + R_{6tc} + R_{ttc2(-40^\circ)}} = \frac{\Delta U_{otc}}{A_{LPFtc}} \tag{6.21}$$

In the formula, $V_{reftc} = 5$ V is the voltage reference of the compensation circuit; δ_{Uotc} is the zero partial temperature variation after scale factor compensation (about 30 mV). $A_{LPFtc} =$ -5 is the amplitude gain of low-pass filter; R_{ttc2} (-40) and R_{ttc2} (60) are the resistance values of thermistor R_{ttc2} (same as R_T) at -40 and 60 °C, respectively. Since the module has no effect on other parts of the circuit, the V_{outtc} temperature trend can be simulated by combining the data in Fig. 6.29, and the zero offset output can be reduced by properly adjusting the zero offset compensation circuit parameters. The zero-bias variation curve of the whole temperature measured after the replacement according to the resistance values in Table 6.4 is

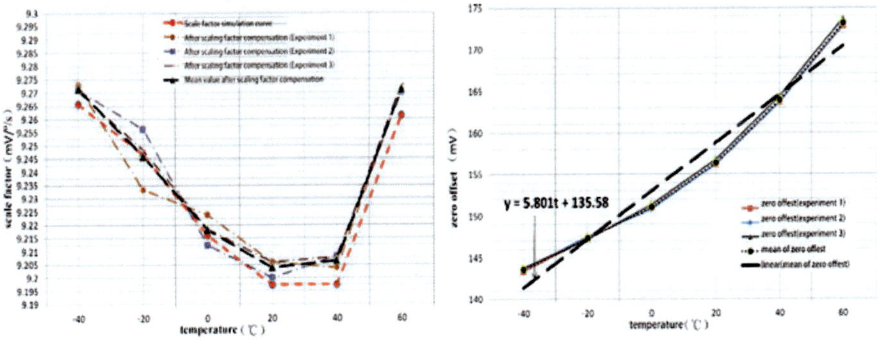

Fig. 6.29 Full temperature test curve and zero deviation test curve after temperature compensation of scale factor

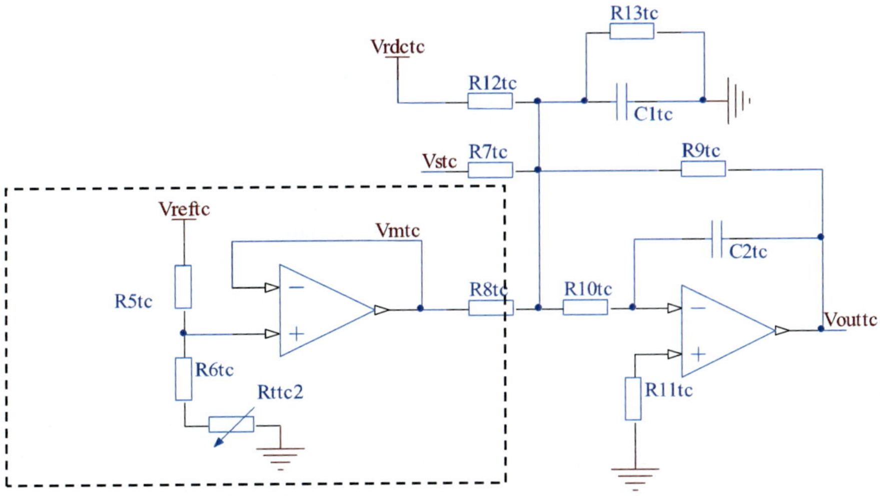

Fig. 6.30 F_{LPF1} circuit diagram

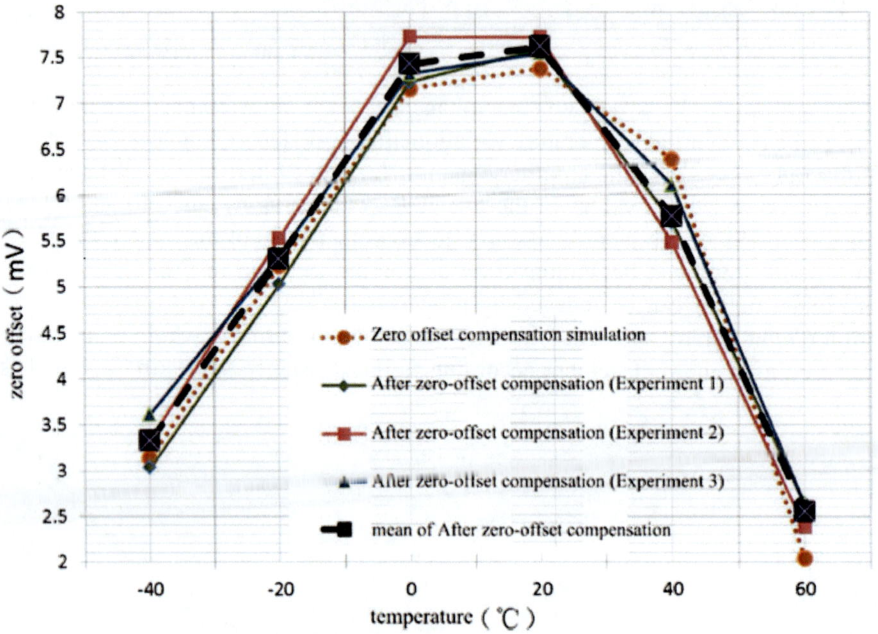

Fig. 6.31 Test results after zero offset temperature compensation

shown in Fig. 6.31. The measured curve basically conforms to the simulation trend. The zero-bias temperature coefficient is about 20°/h/°C (KS = 9.20 mV/o/s), and the compensation effect is obvious. Three groups of repeated experimental curves can be seen from the zero offset compensation repeatability is good.

6.4.2 Temperature Compensation Technology Based on Silicon-Based MEMS Gyroscope Natural Frequency Information

Since the temperature field is not uniform when the ambient temperature is transmitted to the gyroscope, and the calorific value of each part of the measurement and control system circuit is not equal, these factors lead to differences in the temperature distribution in the gyroscope shell, so the position of the temperature sensor has a great influence on the temperature compensation results of the gyroscope. It can be seen from Figs. 6.4, 6.5, 6.6 and 6.7 that the resonant frequency of the silicon-based MEMS gyroscope structure is one-to-one corresponding to the temperature, and it is monotonous. Moreover, the resonant frequency of the gyroscope structure directly reflects the temperature of the structure. Therefore, the current temperature of the gyroscope structure can be compensated by the resonant (inherent) frequency of the gyroscope structure. Compared with the temperature compensation scheme based on temperature sensor, this method has the following advantages: no external temperature detection device is needed, which reduces the circuit cost and reliability; can directly reflect the temperature of the silicon-based MEMS gyroscope structure; the distribution position of temperature detection points does not need to be considered. However, if the analog circuit is used in the measurement and control system, it is necessary to design a high-precision frequency detection circuit, so that the complexity of the circuit is greatly improved. Therefore, the temperature compensation technology based on the natural frequency information of the silicon-based MEMS gyroscope is more suitable for application in the digital measurement and control system.

After the closed loop control of the driving mode of the silicon-based MEMS gyroscope, the driving signal frequency is consistent with the resonant frequency of the driving mode, and the driving sensing voltage amplitude is fixed at the set value. When the temperature changes, under the negative feedback closed loop control, the driving signal frequency and the driving AC voltage amplitude will change with the temperature. At the temperature range of −40–60 °C, the variation of driving signal frequency and driving AC voltage amplitude with temperature is shown in Figs. 6.32 and 6.33. The fitting curves of Figs. 6.32 and 6.33 show that the linearity of driving signal frequency and temperature is much better than that of driving AC voltage amplitude and temperature, so the driving signal frequency can be used as the reference of temperature compensation. Similarly, since the compensation of scale factor will have an impact on zero bias, the temperature compensation of zero bias should be carried out after the compensation of scale factor.

Firstly, the scale factor temperature characteristics of the dual-mass linear vibration silicon-based MEMS gyroscope prototype are tested. The results show that the scale factor temperature coefficient is 254 ppm/°C as shown in Fig. 6.34. It can be seen that the prototype is different from the prototype in the previous section, and its scale factor curve with temperature is not suitable for linear fitting, so piecewise compensation is needed. According to the relationship between temperature and scale factor, the curve was divided into five sections with 0, 20, 40 and 50 °C as the boundary points. With the frequency corresponding to temperature as the independent variable, the scale factor and frequency of each segment are linearly fitted, and the linear fitting equation of frequency and scale factor is obtained. Taking the driving signal frequency as the independent variable into the fitting curve, the corresponding scale factor fitting value is obtained. This value removes the detected modal signal sampled by AD, and the scale factor calculated is the normalized scale factor after compensation. The temperature coefficient of the scale factor after compensation is 9 ppm/°C (as shown in the right Fig. 6.34), which is 1/28 before compensation.

Under the premise of the scale factor temperature compensation system, the temperature compensation is carried out for the zero bias. The temperature experiment shows that the zero bias of the silicon-based MEMS gyroscope decreases with the increase of temperature, and changes by 1.4 mV in the whole temperature range. The zero bias temperature coefficient is 50.4 (°/h)/°C, and the stability of the whole temperature zero bias is 1808.3°/h. The zero bias temperature characteristics are poor. In view of the good effect of segmented compensation of the total

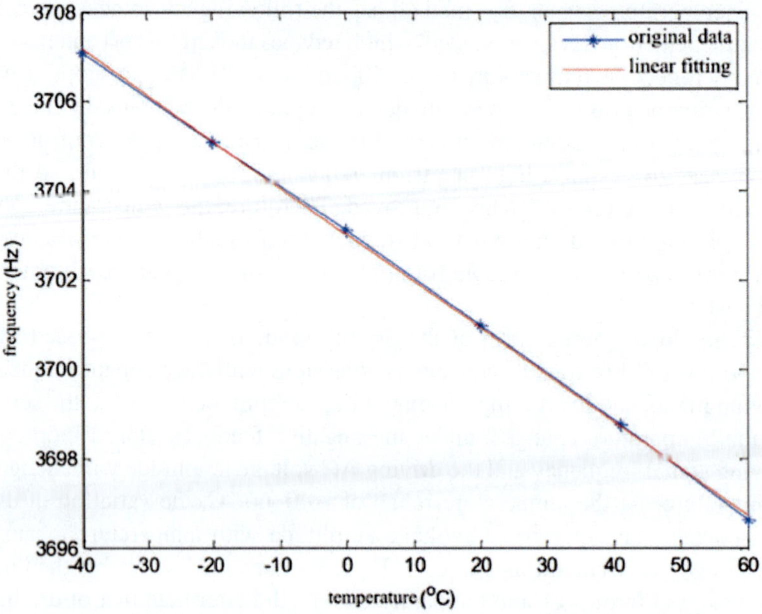

Fig. 6.32 Driving signal frequency versus temperature curve

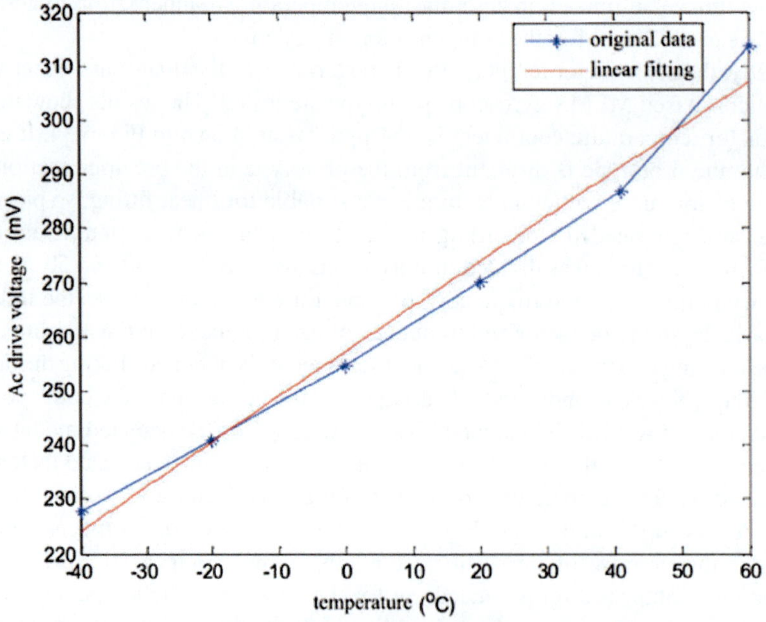

Fig. 6.33 Curve of driving AC voltage amplitude versus temperature

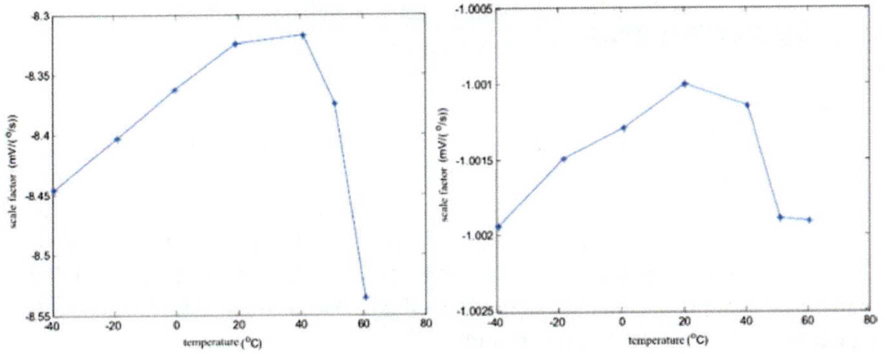

Fig. 6.34 Scale factor temperature curves of a dual-mass linear vibration silicon-based MEMS gyroscope prototype before and after temperature compensation

temperature scale factor, the zero bias change caused by temperature is also compensated by segmented method. According to the relationship between zero deviation and temperature, the temperature of 0, 20, 40 and 50 °C are still divided into five sections as the boundary points. With the corresponding frequency as the independent variable, the zero deviation curves are linearly fitted to obtain the linear fitting equation of frequency and zero deviation. In the actual compensation, the driving signal frequency is substituted into the fitting curve to obtain the fitted zero bias value; by subtracting the zero-offset signal from the demodulated output, the trend of zero offset changing with temperature can be basically eliminated. After the zero bias compensation algorithm is added, the zero bias of the whole temperature is tested according to the same method, and the results are shown in Fig. 6.35. The trend of the compensated zero-bias signal with temperature change is greatly weakened. The zero-bias variation in the whole temperature range is 0.2 mV, the temperature coefficient of zero-bias is reduced to 7.2 (°/h)/°C, and the stability of zero-bias at whole temperature is reduced to 98.02°/h.

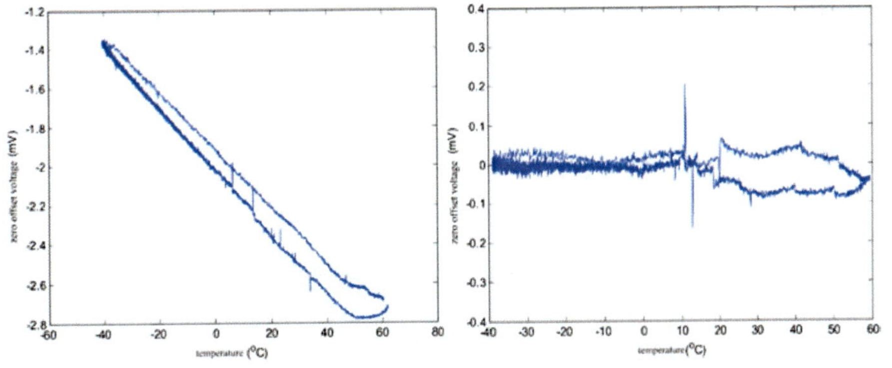

Fig. 6.35 Zero bias temperature curves of a dual-mass linear vibration silicon-based MEMS gyroscope prototype before (left) and after (right) temperature compensation

6.5 Chapter Summary

This chapter first introduces the model of the silicon-based MEMS gyroscope structure affected by temperature change, and analyzes and experimentally verifies it from two aspects of resonant frequency and quality factor. Secondly, the temperature control method is mainly introduced to reduce the influence of external temperature change on the gyroscope. This part is carried out from the chip-level temperature control technology and the whole table-level temperature control technology. Finally, based on the thermal resistance and silicon-based MEMS gyroscope natural frequency, the temperature compensation technology is introduced to alleviate the problem that the scale factor and zero offset are greatly affected by temperature, and the detailed calculation method is given for readers to apply.

Chapter 7
Dual-Mass Linear Vibration Silicon Based MEMS Gyroscope Monitoring Circuit Design and Test Technology

7.1 Chapter Introduction

In the form of analog circuit according to the analysis and related parameters of the drive circuit, sensing interface, quadraturely coupled stiffness correction circuit and PIPLC sensing closed loop circuit controller on the printed circuit board. Since the relevant signal in the system is weak, this chapter refers to the technology of improving the system reliability and jamming resistance in the circuit design and welding process [151–153]. In the dual-mass linear vibration silicon-based MEMS gyroscope "sensing quadrature" state (no quadrature correction, sensing open loop), "sensing quadrature and closed" state (quadrature correction sensing open loop) and "sensing quadrature closed" state (quadrature correction sensing closed loop), and the experimental results are analyzed and compared [154, 155]. Among them, some experimental data have been introduced in the previous article.

7.2 Dual-Mass Linear Vibration Silicon Based MEMS Gyroscope Monitoring Circuit Design

The whole system of measurement and control circuit of dual-mass linear vibration silicon-based MEMS gyroscope consists of three sub-circuits, namely, drive circuit, sensing loop and quadrature correction loop. The block diagram is shown in Fig. 7.1. The whole system consists of three PCB circuit boards, gyro head and drive circuit and interface circuit are installed on board I, board II is quadrature nulling module circuit, and board III is sensing circuit. The three circuit boards are connected by the highly reliable coupling pins, and the impact-resistant rubber pad is placed in the aluminum alloy shell. The physical photos of each part are shown in Fig. 7.2. The overall volume is 46 mm × 46 mm × 36 mm. The circuit is powered by ±8 V power supply, and the power consumption is about 560 mW. Each circuit has been analyzed in detail before, and will not be repeated in this section.

H. Cao, *Dual-Mass Linear Vibration Silicon-Based MEMS Gyroscope*,
https://doi.org/10.1007/978-981-19-9247-6_7

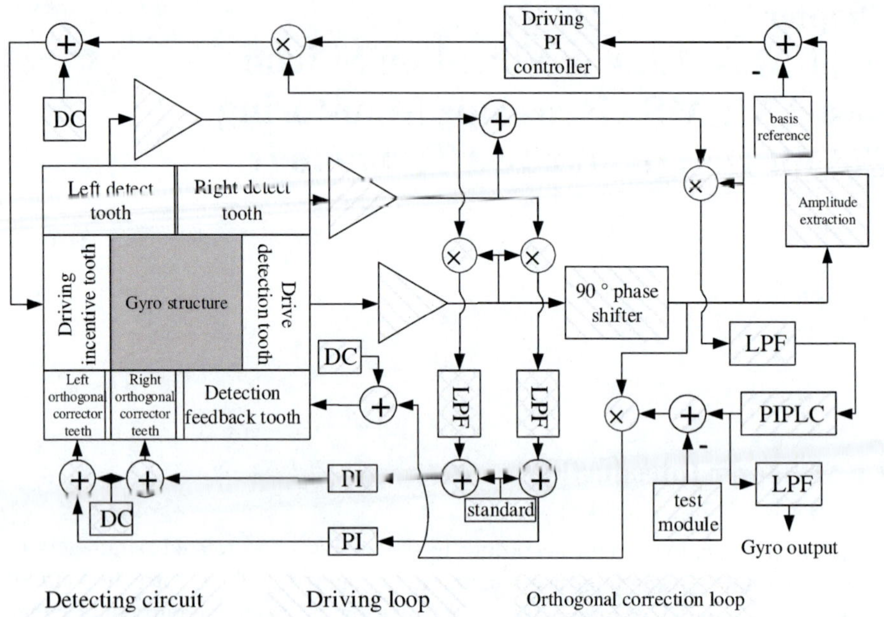

Fig. 7.1 The overall block diagram

Fig. 7.2 Physical photo of the GY-027 gyroscope

7.3 Key Performance Test

In this section, on the basis of existing equipment in the laboratory, dual-mass linear vibration silicon-based MEMS gyroscope prototype GY-027 gyroscope was tested for its fully open loop state, open loop state sensed after quadrature correction and closed loop state sensed after quadrature correction. The experimental equipment is shown in Fig. 7.3, mainly including power supply, temperature control transfer table, digital multimeter, signal source, digital oscilloscope, spectrum analyzer, data acquisition computer, etc.

Fig. 7.3 Physical drawing of the experimental equipment

7.3.1 *Scaling Factor Correlation Index Test*

The GY-027 gyroscope is fixed on the test table so that its input axis points to the sky. After powering on the gyroscope and preheating for 30 min, the console operates at ±0.1°/s, ±0.2°/s, ±0.5°/s, ±1°/s, ±10°/s, ±20°/s, ±100°/s, ±200°/s. The collected data is recorded separately when the output of the gyroscope is stable. Zero-bit data were sampled for 30 s before and after the angular rate test. After repeating the above experiment process three times, the obtained data are shown in Table 7.1.

7.3.1.1 Scale Factor

The mean value of silicon-based MEMS gyroscope output value in group m is:

$$F_j = \frac{1}{N_{test}} \sum_{m=1}^{N_{test}} F_{jm} \tag{7.1}$$

In formula, F_{jm} Output value of the top group m; N_{test} For the number of samples. The linear model of establishing the gyroscope input and output relationship is as follows:

$$F_j = K_n \Omega_{ij} + F_0 + \nu_j \tag{7.2}$$

In formula, K_n Scaling factor, measured in mV/°/s; Ω_{ij} for the input angular speed; F_0 And ν_j for the fit null position and the fit error. The data in Table 7.1 were processed and the scaling factor fitting curves for the three states were plotted as shown in Fig. 7.4, and the relevant scaling factor values are shown in Table 7.2. The scaling factor for sensing closed loop state in Fig. 7.4 is opposite to the other two states as a result of the action of an output-level low-pass filter.

Table 7.1 GY-027 Normal temperature scaling factor experiment

Ω_z(°/s)	Sensing open quadrature open output (V)			"Sensing open quadrature closed" output (V)			"Sensing closed quadrature closed" output (V)		
	First time	Second time	Third time	First time	Second time	Third time	First time	Second time	Third time
200	−2.5545	−2.5535	−2.5536	−1.9897	−1.9895	−1.9920	1.8480	1.8480	1.8481
100	−1.2953	−1.2960	−1.2971	−0.9902	−0.9948	−0.9909	0.9184	0.9186	0.9181
50	−0.6658	−0.6676	−0.6692	−0.4891	−0.4895	−0.4897	0.4534	0.4535	0.4537
20	−0.2886	−0.2907	−0.2927	−0.1870	−0.1871	−0.1871	0.1746	0.1746	0.1747
10	−0.1629	−0.1654	−0.1675	−0.0865	−0.0865	−0.0864	0.0818	0.0817	0.0817
5	−0.0991	−0.1025	−0.1046	−0.0362	−0.0364	−0.0362	0.0352	0.0352	0.0353
2	−0.0622	−0.0646	−0.0669	−0.0061	−0.0062	−0.0061	0.0073	0.0073	0.0074
1	−0.0497	−0.0521	−0.0544	0.0040	0.0038	0.0039	−0.0020	−0.0020	−0.0019
0.5	−0.0433	−0.0458	−0.0481	0.0091	0.0090	0.0090	−0.0066	−0.0066	−0.0066
0.2	−0.0394	−0.0420	−0.0445	0.0121	0.0120	0.0121	−0.0094	−0.0093	−0.0093
0.1	−0.0382	−0.0408	−0.0433	0.0131	0.0130	0.0130	−0.0102	−0.0104	−0.0103
0	−0.0369	−0.0395	−0.0421	0.0141	0.0140	0.0139	−0.0110	−0.0112	−0.0114
−0.1	−0.0357	−0.0383	−0.0408	0.0151	0.0151	0.0150	−0.0122	−0.0123	−0.0121
−0.2	−0.0344	−0.0371	−0.0393	0.0162	0.0161	0.0161	−0.0132	−0.0132	−0.0131
−0.5	−0.0307	−0.0333	−0.0356	0.0191	0.0190	0.0191	−0.0159	−0.0159	−0.0158
−1	−0.0245	−0.0269	−0.0293	0.0241	0.0240	0.0240	−0.0206	−0.0205	−0.0205
−2	−0.0119	−0.0144	−0.0167	0.0341	0.0340	0.0341	−0.0299	−0.0299	−0.0298
−5	0.0256	0.0231	0.0211	0.0642	0.0641	0.0642	−0.0578	−0.0578	−0.0577
−10	0.0886	0.0860	0.0839	0.1143	0.1144	0.1144	−0.1043	−0.1043	−0.1041
−20	0.2140	0.2113	0.2099	0.2147	0.2148	0.2148	−0.1972	−0.1973	−0.1971

(continued)

Table 7.1 (continued)

Ω_z(°/s)	Sensing open quadrature open output (V)			"Sensing open quadrature closed" output (V)			"Sensing closed quadrature closed" output (V)		
	First time	Second time	Third time	First time	Second time	Third time	First time	Second time	Third time
−50	0.5904	0.5877	0.5860	0.5153	0.5154	0.5154	−0.4761	−0.4762	−0.4760
−100	1.2177	1.2141	1.2125	1.0144	1.0149	1.0151	−0.9407	−0.9408	−0.9408
−200	2.4693	2.4646	2.4619	2.0204	2.0197	2.0144	−1.8707	−1.8709	−1.8706

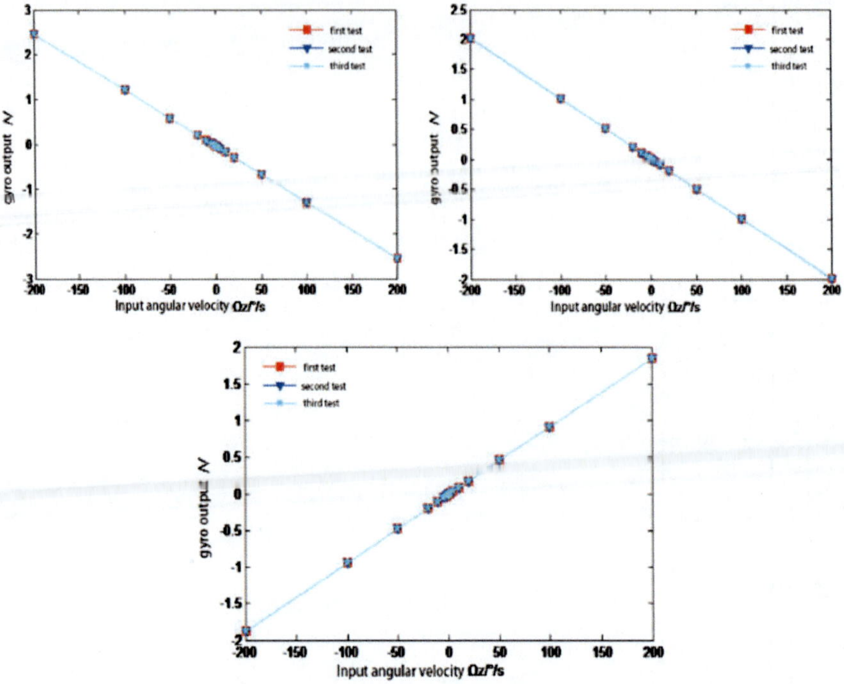

Fig. 7.4 Scale factor test curves of sensing and quadrature both open (top left), sensing open and quadrature closed (top right) and sensing and quadrature both closed (bottom)

7.3.1.2 Scaling Factor of Non-linearity

The nonlinearity of the scale factor can be obtained by using the following expressions:

$$K_m^{'} = \left. \frac{\overline{F}_j - F_j}{K_n|\Omega_{\max+} - \Omega_{\max-}|} \right|_{\max} \tag{7.3}$$

$$K^{'} = \frac{1}{Q_{test}} \sum_{m=1}^{Q_{test}} K_m^{'} \tag{7.4}$$

In the formula, $K_m^{'}$ represents the scaling factor nonlinearity obtained from the m times test; $\overline{F}_j$ represents the j times input angular velocity corresponds to the calculated value on the fitted, Ω_{max+} and Ω_{max-} represent the maximum input angular velocity at forward and reverse, respectively; K' represents the nonlinearity of the scaling factor; Q_{test} represents the number of tests performed.

Table 7.2 GY-027 results of scale factor performance test

State	Number of times	Forward transfer scaling factor (mV/°/s)	Reverse the scaling factor (mV/°/s)	Scale factor (mV/°/s)	Nonlinearity (ppm)	Degree of asymmetry (ppm)	Repetitiveness (ppm)
Sensing and quadrature both open	First time	−12.5874	−12.5341	−12.5607	1031	4219	818
	Second time	−12.5688	−12.5233	−12.5467	923	3666	
	Third time	−12.5573	−12.5227	−12.5413	822	2711	
	Mean			−12.5496	925	3532	
Sensing open and quadrature closed	First	−10.0234	−10.0268	−10.0260	511	398	403
	Second time	−10.0296	−10.0251	−10.0294	659	499	
	Third time	−10.0339	−10.0034	−10.0206	811	392	
	Mean			−10.0253	660	430	
Sensing and quadrature both closed	First	9.29612	9.29674	9.29662	69.87	66.69	50.36
	Second time	9.29673	9.29745	9.29718	39.04	76.37	
	Third time	9.29616	9.29657	9.29625	68.90	44.10	
	Mean			9.29668	59.27	62.39	

7.3.1.3 Scaling Factor Asymmetry Degree

The scale factor asymmetry of the silicon-based MEMS gyroscope has the calculation formula of:

$$K_{mu} = \frac{|K_{+} - K_{-}|}{\overline{K}} \tag{7.5}$$

$$K_{u} = \frac{1}{Q}\sum_{m=1}^{Q} K_{mu} \tag{7.6}$$

In the formula, K_{mu} represents the scaling factor asymmetry corresponding to the m times test; K_{+} and K_{-} represent the scale factor obtained in the positive and negative input angular velocity range respectively; $\overline{K}$ represents the average scale factor; K_{u} represents the scale factor asymmetry. The scaling factor asymmetry indexes obtained from Table 7.1 are shown in Table 7.2.

7.3.1.4 Scaling Factor Repeatability

Repeat the scaling factor measurement Q_{test} times, the silicon-based MEMS gyroscope and test circuit must be turned off for a period of time to reach room temperature. An expression for the reproducibility of the scale factor can be expressed as:

$$K_r = \frac{1}{\overline{K}}\sqrt{\frac{1}{Q_{test}-1}\sum_{i=1}^{Q_{test}}(K_i-\overline{K})^2} \tag{7.7}$$

In formula, K_r represents scale factor repeatability; $\overline{K}$ represents the average scale factor; K_i represents the scale factor of the first measurement. The reproducible measures of the scale factors are shown in Table 7.2.

7.3.1.5 Threshold

After measuring the scale factor related performance to maintain the gyroscope electric state, first measure the input angular rate of the gyroscope output of zero, and then apply a specified angular rate input to the gyroscope, with 1s sampling period of 10 s, record the gyroscope output, the gyroscope output relative to the zero rate input change should be greater than 50% of the output value corresponding to the scale factor. The above steps were repeated in the opposite angular rate input direction to obtain the reverse direction threshold data for three consecutive trials. The maximum value of the minimum angular rate input value in two directions is the gyroscope threshold. The test threshold of the GY-027 gyroscope is 0.003°/s, and the test data are shown in Table 7.3.

7.3.1.6 Resolution Ratio

Continue to keep the gyroscope electric, first to the gyroscope with a threshold of 30 times the input angular rate, turn stable and then add a specified rate increment, the gyroscope output increment should be greater than 50% of the corresponding scale factor output increment, then back to the original normal input rate, then reduce a specified rate value, measuring the gyroscope output transformation should be greater than the scale factor of the corresponding output change of 50%. The above tests were repeated in the opposite injection direction to obtain the resolution data in the opposite direction, and 3 trials were repeated. The minimum rate increment maximum is the gyroscope resolution in both directions. Table 7.4 shows the resolution test data with a resolution of 0.003°/s.

Table 7.3 Results of the GY-027 threshold testing

	0.003°/s first time (mV)		0.003°/s second time (mV)		0.003°/s third time (mV)		
	Corotation	Reversal	Corotation	Reversal	Corotation	Reversal	Zero input(mV)
Point one	−11.0615	−11.1277	−11.1023	−11.1382	−11.0581	−11.1173	
Point two	−11.1609	−11.0713	−11.0337	−11.0720	−11.0532	−11.0405	
Point three	−11.1488	−11.1219	−11.0371	−11.0160	−10.9718	−11.1091	
Point four	−11.1229	−11.1617	−11.0482	−11.1380	−10.9822	−11.0702	
Point five	−11.1194	−11.0895	−11.0884	−11.1324	−11.0194	−11.0823	
Point six	−11.0978	−11.1275	−11.0475	−11.0867	−11.0787	−11.0843	
Point seven	−10.9917	−11.1226	−11.0061	−11.1099	−11.0317	−11.1203	
Point eight	−11.0213	−11.1097	−11.1117	−11.1192	−11.0498	−11.1146	
Point nine	−11.0711	−11.0875	−11.0701	−11.1100	−11.0909	−11.1019	
Point one	−11.0834	−11.0736	−11.0293	−11.0823	−11.0862	−11.0983	
Mean	−11.0908	−11.1073	−11.0524	−11.0963	−11.0388	−11.0976	−11.0727

Table 7.4 GY-027 resolution test results

	First (mV)		Second time (mV)		Third time (mV)		
	0.087°/s	0.093°/s	0.087°/s	0.093°/s	0.087°/s	0.093°/s	0.09°/s (mV)
Point one	−10.3511	−10.2305	−10.2723	−10.1921	−10.3073	−10.1572	
Point two	−10.2397	−10.1694	−10.2862	−10.1895	−10.2103	−10.1608	
Point three	−10.2092	−10.1701	−10.2175	−10.2262	−10.2026	−10.2537	
Point four	−10.3100	−10.1788	−10.2562	−10.1958	−10.2107	−10.2781	
Point five	−10.2282	−10.2180	−10.2772	−10.1941	−10.2524	−10.1696	
point six	−10.2707	−10.1880	−10.2557	−10.2543	−10.2535	−10.1505	
Point seven	−10.3042	−10.2227	−10.2646	−10.2492	−10.3711	−10.1347	
Point eight	−10.2277	−10.2656	−10.2314	−10.0986	−10.2407	−10.2202	
Point nine	−10.2521	−10.2659	−10.2491	−10.1209	−10.2581	−10.2306	
Point one	−10.2681	−10.1607	−10.3179	−10.2444	−10.2918	−10.192	
Mean	−10.2661	−10.207	−10.2628	−10.1965	−10.2599	−10.1989	−10.2367

7.3.1.7 Scaling Factor Temperature Factor

The gyroscope is powered up, and the temperature control box of the temperature control rate turntable is increased to 60 °C at the maximum temperature rise rate. After one hour of insulation, the turntable is ordered to turn at ±1°/s rate for 30 s to record the data, and the scaling factor corresponding to 60 °C is obtained. Reduce the maximum cooling rate to −40 °C and heat preservation for one hour, and the turntable is allowed to rotate at ±1°/s rate for 30 s to record the data, and obtain the scaling factor corresponding to −40 °C. The scaling factor temperature coefficient K_{TE} is calculated according to the following equation:

$$K_T = \frac{F_{(+1^\circ/s)} - F_{(-1^\circ/s)}}{2} \tag{7.8}$$

$$K_{TE} = \left|\frac{K_T - K_n}{100K_n}\right|_{max} \times 10^6 \tag{7.9}$$

Among, K_T represents the scaling factor at a certain temperature; F $_{(+1°/s)}$ and F $_{(-1°/s)}$ represent the top output value when the turntable rotates at ±1°/s at this temperature; K_n represents the scaling factor of the normal temperature gyroscope. Test data of full temperature scale factor performance of GY-027 gyroscope are shown in the following Table 7.5.

Table 7.5 GY-027 full temperature scaling factor performance test results

State	Number of times	+1°/s mean (mV)	−1°/s mean (mV)	Scale factor (mV/°/s)	Scaling factor temperature factor (ppm/°C)
Sensing open and quadrature open	60 °C	114.9052	141.5788	−13.3318	623
	−40 °C	90.1422	115.2191	−12.5385	
	Normal atmospheric temperature	–	–	−12.5496	
Sensing open and quadrature closed	60 °C	−22.0177	−1.6055	−10.2061	180
	−40 °C	−23.9197	−4.16233	−9.87869	
	Normal atmospheric temperature	–	–	−10.0253	
Sensing and quadrature both closed	60 °C	−1.6302	−20.1909	9.28037	28.71
	−40 °C	−2.4397	−20.9797	9.26999	
	Normal atmospheric temperature	–	–	9.29668	

7.3.2 Bias Correlation Performance Test

When the gyroscope is fixed on the test table, set the computer sampling period as 1s. After checking that the equipment is connected correctly, the sampling record is started for 80 min and the valid data for 60 min. The test is repeated 7 times for one hour. Seven sets of zero-biased output curves for each state are shown in Figures 7.5, 7.6 and 7.7.

7.3.2.1 Zero Offset

Zero-biased data is available according to the following formula:

$$\overline{F} = \frac{1}{N_{test}} \sum F_i \tag{7.10}$$

$$B_{0m} = \frac{1}{K_n} \overline{F} \tag{7.11}$$

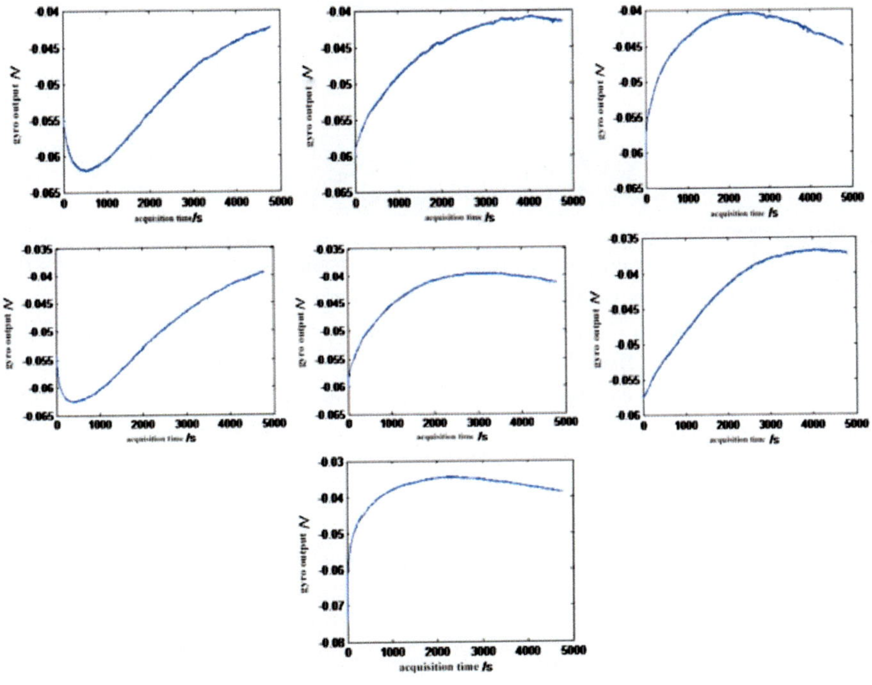

Fig. 7.5 Seven-fold zero-bias output curve of "sensing and quadrature both open" state

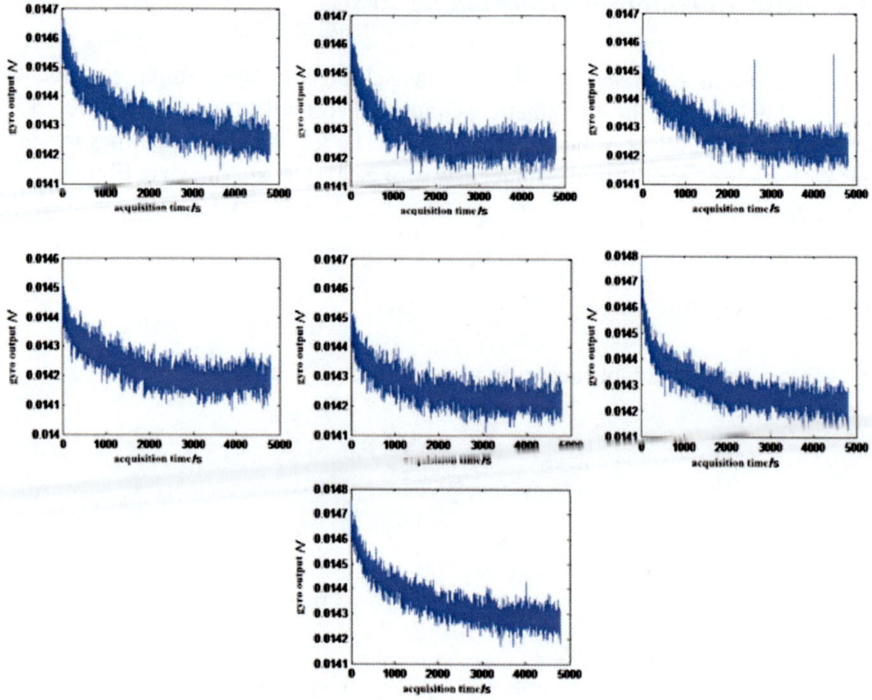

Fig. 7.6 Seven-fold zero-bias output curve of "sensing open and quadrature closed" state

$$B_0 = \frac{1}{Q_{test}} \sum_{m=1}^{Q_{test}} B_{0m} \tag{7.12}$$

In formula, F_i represents the voltage output for each sample; $\overline{F}$ represents the average output obtained from N tests; B_{0m} represents the Gyro zero bias obtained for the m times test; B_0 represents the gyro zero deviation. According to the curve in the figure, the zero-offset drift of "sensing quadrature" state is very obvious and has a large range. The drift range is more than 20 mV within 80 min of sampling. According to the scaling factor given in Table 7.2, the drift is about 2°/s. After adding the quadrature nulling system, the zero-offset drift time and range are greatly improved, but the start time after power supply is longer, and the smooth output can be achieved after about 2000 s. There is no significant large range drift in the system after sensing closed loop, which greatly reduces the start time.

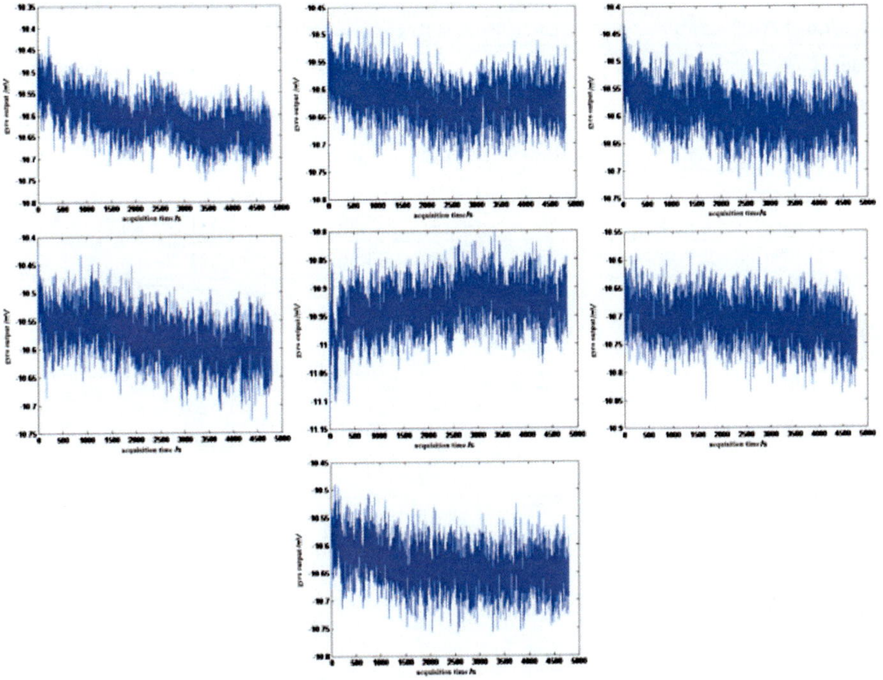

Fig. 7.7 Seven-fold zero-bias output curve of "sensing closed quadrature closed" state

7.3.2.2 Zero Partial Stability

The expression for the zero-bias stability can be expressed as:

$$B_{sm} = \frac{1}{K_n}\left[\frac{1}{(N/P-1)}\sum_{j=1}^{N/P}(F_j-\overline{F})^2\right]^{0.5} \tag{7.13}$$

$$B_s = \frac{1}{Q_{test}}\sum_{m=1}^{Q_{test}} B_{sm} \tag{7.14}$$

In formula, F_j represents the new data sample obtained by taking the average value by period P; P = 10 is the data average period; B_{sm} represents the zero bias stability for the m times test; B_s represents the zero-bias stability.

Table 7.6 GY-027 zero bias performance test results at room temperature

State	Sensing and quadrature both open		Sensing open and quadrature losed		Sensing and quadrature both closed	
Test number of times	Zero offset (°/s)	Zero partial stability (°/h)	Zero offset (°/s)	Zero partial stability (°/h)	Zero offset (°/s)	Zero partial stability (°/h)
First	−5.492	890	1.747	25.88	−1.521	8.31
Second time	−4.795	563	1.752	24.56	−1.523	6.27
Third time	−4.586	374	1.748	23.37	−1.52	7.45
The fourth time	−4.549	309	1.741	20.15	−1.517	8.73
The fifth time	−5.518	482	1.744	18.06	−1.528	6.42
The sixth time	−4.450	812	1.749	27.67	−1.526	5.97
The seventh time	−4.015	348	1.756	28.65	−1.526	6.67
Average	−4.772	540	1.748	24.05	−1.523	7.12
Repetitiveness	1991°/h		17.81°/h		14.10°/h	

7.3.2.3 Zero Partial Repeatability

An expression for zero-biased reproducibility can be expressed as:

$$B_r = \left[\frac{1}{Q_{test} - 1} \sum_{m=1}^{Q_{test}} (B_{0m} - \overline{B}_0)^2 \right]^{0.5} \tag{7.15}$$

In formula.g$_r$ represents the zero partial repeatability; $\overline{B}_0$ represents the zero partial average. From Table 7.6, the zero bias stability of the three states is 540°/h, 24.05°/h and 7.12°/h, and the zero bias repeatability is 1991°/h, 17.81°/h and 14.10°/h, respectively. Therefore, the top zero and partial performance is greatly improved by quadrature correction and sensing closed loop control, which is consistent with the analysis conclusion of the previous chapter.

7.3.2.4 Zero Partial Temperature Coefficient

Heat the temperature box to 60 °C at the maximum allowable rate of the equipment and keep warm for one hour. Power the gyroscope to work, conduct static testing and sampling for one hour; then cool the temperature box to −40 °C for static testing and sampling during cooling; drop to −40 °C, heat the gyroscope for one hour for static testing and sampling at the maximum rate to 60 °C and conduct static testing and sampling during heating; reach 60 °C and heat the gyroscope for one hour after sampling. Given the zero curve of full temperature, zero partial temperature coefficient:

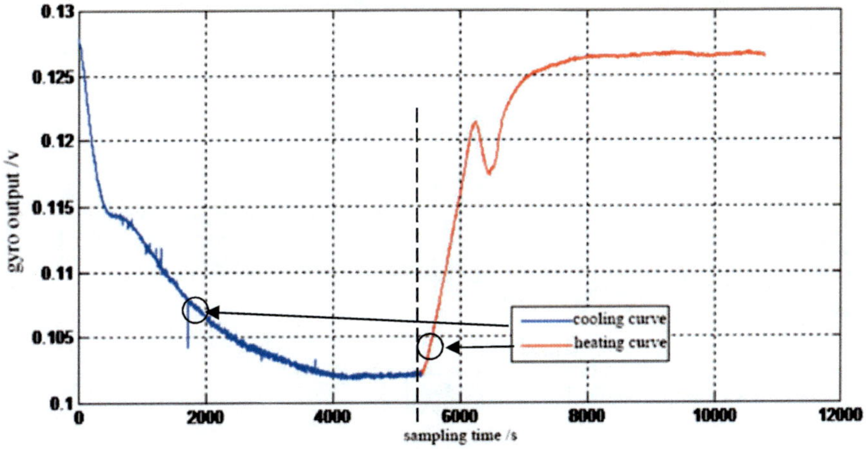

Fig. 7.8 "Sensing and quadrature both open" state zero bias full temperature curve (The left side of the dotted line is the cooling curve, and the right side of the dotted line is the heating curve)

$$B_t = \frac{B_{\max} - B_{\min}}{100} \times 3600 \tag{7.16}$$

In formula. g_{max} and B_{min} represent the Data maximum and minimum, respectively, in ° /s. The full temperature curves of the gyro in the three states are shown in Figs. 7.8, 7.9 and 7.10, respectively. In each figure, the noise in the cooling process is larger than the heating process, which is that additional noise occurs by the high power compressor during the cooling. The zero-biased temperature coefficients for each state are shown in Table 7.7.

7.4 Overall Performance Index Summary

Summarizing the above test results, the performance parameters of GY-027 gyroscope selected in this book are shown in Table 7.7. Therefore, it can be seen from the table that the gyroscope parameters are significantly improved under the action of quadrature correction technology and sensing closed loop technology. It should be noted that the range to sensing open loop state is smaller than the closed loop state because the sensing mode displacement to sense the open loop state when the high speed input is too large (especially in the driveing anti-phase and sensing anti-phase mode frequency difference is small).At the same time, Eq. (3.23) shows that the large motion of the sensing mode will add a non-negligible Coriolis force $2m_c\Omega_z\dot{y}$ to the driving mode, which will be coupled to the driving circuit, resulting in circuit instability and large nonlinear error of the scaling factor. However, the sensing mode displacement of the sensing closed loop state is almost zero, and the large input

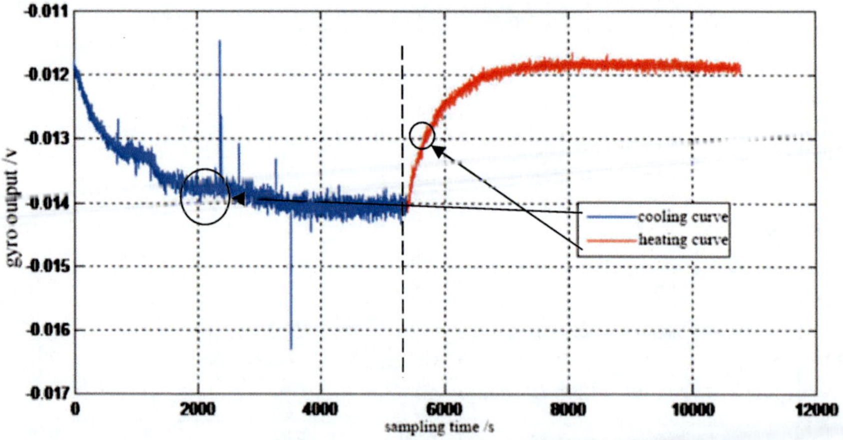

Fig. 7.9 "Sensing open quadrature closed" state zero bias full temperature curve (The left side of the dotted line is the cooling curve, and the right side of the dotted line is the heating curve)

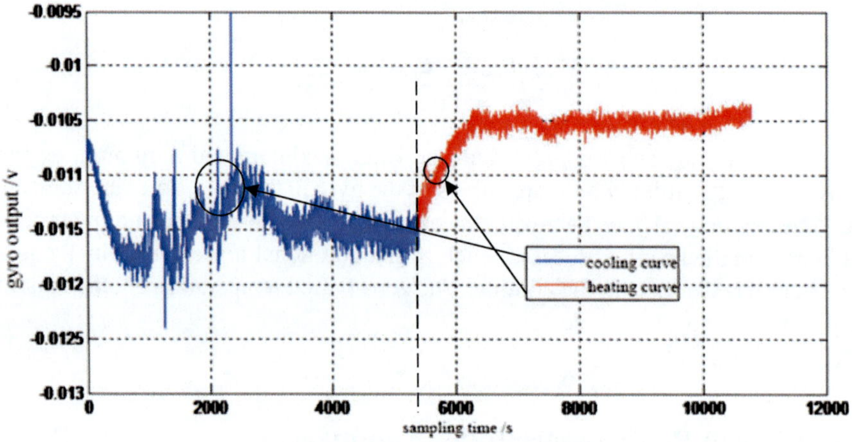

Fig. 7.10 "Sensing and quadrature both closed" state zero bias full temperature curve (The left side of the dotted line is the cooling curve, and the right side of the dotted line is the heating curve)

Table 7.7 GY-027 full temperature zero bias performance test results

State	B_{max} (mV)	B_{min} (mV)	Zero-partial temperature coefficient, (°/h/°C)
Sensing and quadrature both open	127.7353	101.8993	74.11
Sensing open and quadrature losed	−11.3115	−14.2607	10.59
Sensing and quadrature both closed	−10.4002	−11.3272	3.59

angular rate can be responded to because the large sensing mode displacement is negligible. The sensing closed loop based on analog circuit studied in this book is longitudinally compared with the previous work content on the basis of no temperature compensation link. All parameters are shown in Table 7.8. The table shows that the zero-bias stability, zero-bias temperature coefficient, bandwidth and other indicators are better than the previous work, reflecting the significance of the book work (Table 7.9).

Table 7.8 Horizontal comparison of the performance indexes of GY-027 gyroscope

		Gyro state		
Test item	Dimension	Sensing and quadrature both open	Sensing open and quadrature closed	Sensing and quadrature both closed
Scale factor	mV/°/s	−12.5496	−10.0253	9.29668
Scaling factor of non-linearity	ppm	925	660	59.27
Scaling factor asymmetry degree	ppm	3532	430	62.39
Scaling factor repeatability	ppm	818	403	50.36
Scaling factor temperature factor	ppm/°C	623	180	28.71
Range	°/s	±200	±200	±500
Threshold	°/s	–	–	0.003
Resolution ratio	°/s	–	–	0.003
Start time	s	1	1	1
Zero offset	°/s	−4.772	1.747	−1.521
Zero partial stability	°/h	540	24.05	7.12
Zero partial repeatability	°/h	1991	17.81	14.10
Full temperature, zero bias, and temperature coefficient	°/h/°C	74.11	10.59	3.59
Bandwidth	Hz	10	12	104

Table 7.9 Longitudinal comparison of the performance indicators of the dual-mass linear vibration silicon-based MEMS gyroscope prototype in this book

		Gyro state	
Test item	Dimension	Digital circuit sensing open loop [118] (With temperature compensation)	This book sensing close loop (Without temperature compensation)
Scale factor	mV/°/s	−0.9997	9.29668
Scaling factor of non-linearity	ppm	22.6	59.27
Scaling factor asymmetry degree	ppm	60.58	62.39
Scaling factor repeatability	ppm	59.23	50.36
Scaling factor temperature factor	ppm/°C	45	28.71
Range	°/s	± 200	± 500
Threshold	°/s	0.02	0.003
Resolution ratio	°/s	0.02	0.003
Start time	s	10	1
Zero offset	°/s	0.28	−1.521
Zero partial stability	°/h	27.01	7.12
Zero partial repeatability	°/h	44.96	14.10
Zero partial temperature coefficient	°/h/°C	14.4	3.59
Bandwidth	Hz	–	104

7.5 Chapter Summary

This chapter first introduces the working principle of the whole dual-mass linear vibration silicon-based MEMS gyroscope overall measurement and control system, including the driving closed loop, quadrature correction closed loop and sensing closed loop. Secondly, the three states of sensing open loop, sensing open loop oxy_insert_endquadrature closed loop and sensing closed loop quadrature closed loop were tesed for GY-027gyroscope dual-mass linear vibration silicon-based MEMS gyroscope prototype, and the results showed that the scaling factor and zero bias were greatly improved. The zero bias stability was increased from 540°/h to 24.05°/h after adding the sensing closed loop controller, and the zero bias stability reached 7.12°/h. At the same time, the temperature performance of the gyroscope has been greatly improved (full temperature and zero bias temperature coefficient is 3.59°/h/°C), and even beyond the prototype with temperature compensation function, proving the feasibility and practicability of the research content of this book.

References

1. Wang Shourong, Huang Libin, Yang Bo. Microinertial Instrument and Microsystem [M]. Beijing: Ordnance Industry Press, 2011
2. Yazdi N, Ayazi F, Najafi K. Micromachined inertial sensors[J]. Proceedings of the IEEE, 1998, 86(8): 1640–1659
3. Geen J A. Very low cost gyroscopes [C]. Proc. IEEE Sensors, 2005: 537–540
4. Cao Huiliang, Zhang Yingjie, Kou Zhiwei, et al. Research overview of anti-high overload MEMS gyroscope [J]. Journal of Hebei University of Science and Technology, 2018, 39(4): 289-298
5. Wang Wei, HE Sheng. Development Trend of MEMS Inertial Instrument Technology [J]. Missiles and space Launch vehicles, 2009, 301(3): 23–28
6. Soderkvist J. Micromachined gyroscopes[J]. Sensors and Actuators A, 1994, 43(1) : 65–71
7. Lawrence A. Modern inertial technology: navigation, guidance, and control [M]. New York: springer Verlag, 1998. 21–25
8. Pansiot J, Zhang B, Yang G Z. WISDOM: wheelchair inertial sensors for displacement and orientation monitoring [J] . Meas. Sci. Technol., 2011, 22: 105801
9. Li Xinguang, Yuan Jianping. Development status of micromechanical gyro [J]. Mechanical progress, 2003, 33(3): 289–301
10. Liu Jun, Shi Yunbo, Li Jie. Microinertial technology [M]. Beijing: Publishing House of Electronics Industry, 2005
11. Li Jinming. Design of capacitive micromechanical gyroscope [M]. Beijing: National Defense Industry Press, 2006
12. Sung W T, Lee J Y, Lee J G and Kang T. Design and fabrication of an automatic mode controlled vibratory gyroscope [C]. Proc. IEEE MEMS, 2006. 674–677
13. Sung W T, Sung S, Lee J G and Kang T. Design and performance test of a MEMS vibratory gyroscope with a novel AGC force rebalance control [J]. Journal of Micromechanics and Microengineering, 2007, 17(3): 1939–1948
14. Chang B S, Sung W T, Lee J G, et al. Automatic mode matching control loop design and its application to the mode matched MEMS gyroscope [C]. IEEE International Conference on Vehicular Electronics and Safety, 2007. 1–6
15. Sung S, Sung W T, Kim C, et al. On the mode-matched control of mems vibratory gyroscope via phase-domain analysis and design [J]. IEEE/ASME Transactions on Mechatronics, 2009, 14(4): 446–455
16. Sung W T, Sung S, Lee J Y, et al. Development of a lateral velocity-controlled MEMS vibratory gyroscope and its performance test [J]. Journal of Micromechanics and Microengineering, 2008, 18, 055028
17. Acar C, Schofield A R, Trusov A A, et al. Environmentally robust MEMS vibratory gyroscopes for automotive applications [J]. IEEE Sensors Journal, 2009, 9(12): 1895–1906

H. Cao, *Dual-Mass Linear Vibration Silicon-Based MEMS Gyroscope*,
https://doi.org/10.1007/978-981-19-9247-6

18. Trusov A A. Investigation of factors affecting bias stability and scale factor drifts in Coriolis vibratory MEMS gyroscopes [D]. University of California, Irvine, 2009
19. Trusov A A, Schofield A R, Shkel A M. Performance characterization of a new temperature-robust gain-bandwidth improved MEMS gyroscope operated in air [J]. Sensors and Actuators A: Physical, 2009, 155: 16–22
20. Painter C C, Shkel A M. Active structural error suppression in MEMS vibratory rate integrating gyroscopes [J]. IEEE Sensors Journal, 2003, 3(5): 595–606
21. Prikhodko I P, Zotov S A, Trusov A A, et al. Sub-degree-per-hour silicon MEMS rate sensor with 18million Q-factor [C]. Proc. 16th International Solid-State Sensors, Actuators and Microsystems Conference Transducers, 2011. 2809–2812
22. Trusov A A, Schofield A R, Shkel A M. A substrate energy dissipation mechanism in in-phase and anti-phase micromachined z-axis vibratory gyroscopes [J]. Journal of Micromechanics and Microengineering, 2008, 18(9): 1–10
23. Trusov A A, Schofield A R, Shkel A M. Gyroscope architecture with structurally forced anti-phase drive-mode and linearly coupled anti-phase [C]. Proc. International Solid-State Sensors, Actuators and Microsystems Conference Transducers, 2009. 660–663
24. Trusov A A, Schofield A R, Shkel A M. Study of substrate energy dissipation mechanism in in-phase and anti-phase micromachined vibratory gyroscopes [C]. Proc. IEEE Sensors, 2008. 168–171
25. Trusov A A, Schofield A R, Shkel A M. Micromachined rate gyroscope architecture with ultra-high quality factor and improved mode ordering [J]. Sensors and Actuators A: Physical, 2011, 165: 26–34
26. Fan L S, Tai Y C, Muller R S. Integrated movable micromechanical structures for sensors and actuators [J]. IEEE Transactions on Electron Devices, 1988, 35(6): 724–730
27. Clark W A, Micromachined vibratory rate gyroscopes [D]. University of California, Berkeley, 1997
28. Palaniapan M, Howe R T, Yasaitis J. Performance comparison of integrated z-axis frame microgyroscopes [C]. The IEEE Sixteenth Annual International Conference on Micro Electro Mechanical Systems, 2003. 482–485
29. Yoon S W, Lee S, Perkins N C and Najafi K. Analysis and wafer-level design of a high-order silicon vibration isolator for resonating MEMS devices[J]. Journal of Micromechanics and Microengineering, 2011, 17: 015017
30. Keymeulen D, Fink W, Ferguson M I, et al. Tuning of MEMS devices using evolutionary computation and open-loop frequency response [C]. IEEE Aerospace Conference Proceedings, 2005. 1–8
31. Shcheglov K, Evans C, Gutierrez R, et al. Temperature dependent characteristics of the JPL silicon MEMS gyroscope [C]. IEEE Aerospace Conference Proceedings, 2000. 403–411
32. Keymeulen D, Peay C, Yee K, et al. Effect of temperature on MEMS vibratory rate gyroscope [C]. IEEE Aerospace Conference, 2005. 1–6
33. Ferguson M I, Keymeulen D, Hayworth K, et al. A hardware platform for tuning of MEMS devices using closed-loop frequency response [C]. IEEE Aerospace Conference, 2005. 1–7
34. Iyer S V. Modeling and simulation of non-idealities in a Z-axis CMOS-MEMS gyroscope [D]. Carnegie Mellon University, 2003
35. Xie H K, Fedder G K. A CMOS-MEMS lateral-axis gyroscope [C]. The 14th IEEE International Conference on Micro Electro Mechanical Systems, 2001. 162–165
36. Xie H K, Fedder G K. A DRIE CMOS-MEMS gyroscope [C]. Sensors Proceedings of IEEE, 2002. 1413–1418
37. Xie H K, Fedder G K. Fabrication, characterization, and analysis of a DRIE CMOS-MEMS gyroscope[J]. IEEE Sensors Journal, 2003, 3(5): 622–631
38. Sharma A, Zaman M F, et al. A 0.1°/hr bias drift electronically matched tuning fork microgyroscope [C]. IEEE 21st International Conference on Micro Electro Mechanical Systems, 2008. 6–9
39. Sharma A, Zaman N F, Ayazi F. A sub-0.2°/hr bias drift micromechanical silicon gyroscope with automatic CMOS mode-matching [J]. Journal of Solid-State Circuits, 2009, 44(5): 1593–1608

40. Zaman M F, Sharma A, Hao Z, et al. A mode-matched silicon-yaw tuning-fork gyroscope with subdegree-per-hour Allan deviation bias instability[J]. Journal of Microelectromechanical Systems, 2008, 17(6): 1526–1536
41. Zaman M F, Sharma A, Ayazi F. High performance matched-mode tuning fork gyroscope [C]. Proc. Int. Conf. MEMS, 2006. 66–69
42. Greiff P, Boxenhorm B, King T, et al. Silicon monolithic micromechanical gyroscope [C]. The 6th International Conference on Solid-State Sensors and Actuators, 1991. 966–968
43. Greiff P, Antkowiak B, Campbell J, et al. Vibrating wheel micromechanical gyro [C]. IEEE Position location and navigation symposium, 1996. 31–37
44. Boxenhorn B, Greiff P. A vibratory micromechanical gyroscope [C]. AIAA Guidance and Controls Conference, 1988
45. Bernstein J, Cho S, King A T, et al. A micromachined comb-drive tuning fork rate gyroscope [J]. IEEE Micro Electro Mechanical Systems (MEMS) Proceedings, 1993: 143–148
46. Johnson B R, Weber M W. MEMS gyroscope with horizontally oriented drive electrodes [P]. United States Patent, US 7036373 B2
47. Johnson B R, Cabuz E, French H B, et al. Development of a MEMS gyroscope for northfinding applications [C]. IEEE/ION Position Location and Navigation Symposium, 2010. 168–170
48. Hanse J G. Honeywell MEMS inertial technology & product status [C]. IEEE Position Location and Navigation Symposium, 2004. 43–48
49. Lautenschlager E J. MEMS inertial technology for aerospace applications [C]. Proceedings of CANEUS, 2006. 11083: 1–6
50. Geiger W J, Bartholomeyczik J, Breng U, et al. MEMS IMU for AHRS applications [C]. IEEE/ION Position, Location and Navigation Symposium, 2008. 225–231
51. Gomez U M, Kuhlmann B, Classen J, et al. New surface micromachined angular rate sensor for vehicle stabilizing systems in automotive applications [C]. The 13th International Conference on Solid-State Sensors, Actuators and Microsystems, 2005. 184–187
52. Neul R, Gómez U, Kehr K, et al. Micromachined gyros for automotive applications [C]. Proceedings of the Fourth IEEE Conference on Sensors, 2005. 527–530
53. Geiger W, Folkmer B, Sobe U, et al. New designs of micromachined vibrating rate gyroscopes with decoupled oscillation modes [J]. Sensors and Actuators A, 1998, 66: 118–124
54. Sharma M, Sarraf E H, Cretu E. Parametric amplification-damping in MEMS gyroscopes [C]. Proc. Int. Conf. MEMS, 2011. 607–620
55. Sharma M, Sarraf E H, Baskaran R, et al. Parametric resonance amplification and damping in MEMS gyroscopes [J]. Sensors and Actuators A: Physical, 2012, 177: 79–86
56. Elsayed M, Nabki F, Sawan M, et al. A 5V MEMS gyroscope with 3aF/°/s sensitivity, 0.6°/$\sqrt{}$hr mechanical noise and drive-Sense crosstalk minimization [C]. 2011 International Conference on Microelectronics (ICM), 2011. 1–5
57. Rajaraman V, Sabageh I, French P, et al. Design, modelling and fabrication of a 40-330Hz dual-mass MEMS gyroscope on thick-SOI technology [J]. Procedia Engineering, 2011, 25: 647–650
58. Tatar E, Alper S E, Akin T, et al. Quadrature-error compensation and corresponding effects on the performance of fully decoupled MEMS gyroscopes[J]. Journal of Microelectromechanical Systems, 2012, 21(3): 656–667
59. Tatar E, Alper S E, Akin T. Effect of quadrature error on the performance of a fully-decoupled MEMS gyroscope [C]. Proc. Int. Conf. MEMS, 2011. 569–572
60. Sonmezoglu S, Alper S E, Akin T. An automatically mode-matched MEMS gyroscope with 50 Hz bandwidth [C]. Proc. Int. Conf. MEMS, 2012. 523–526
61. Saukoski M, Aaltonen L, Halonen K A I. Effects of synchronous demodulation in vibratory MEMS gyroscopes: A theoretical study [J]. IEEE Sensors Journal, 2008, 8(10): 1722–1733
62. Saukoski M, Aaltonen L, Halonen K A I. Zero-rate output and quadrature compensation in vibratory MEMS gyroscopes [J]. IEEE Sensors Journal, 2007, 7(12): 1639–1652
63. Saukoski M, Aaltonen L, Salo T, et al. Readout and control electronics for a microelectromechanical gyroscope [C]. Instrumentation and Measurement Technology Conference, 2006. 1741–1746

64. Gallacher B J, Hedley J, Burdess J S, et al. Electrostatic nulling of structural imperfections present in a microring gyroscope [J]. Journal of Microelectromechanical Systems, 2005, 14(2): 221–234
65. Chaumet B, Leverrier B, Rougeot C, et al. A new silicon tuning fork gyroscope for aerospace applications [J]. Syposium Gyro Technology 2009. 1.1–1.13
66. Antonello R, Oboe R, Prandi L and iganzoli F. Automatic mode matching in mems vibrating gyroscopes using extremum-seeking control [J]. IEEE Transactions on Industrial Electronics, 2009, 56(10): 3880–3891
67. Khankhua S, Ashraf M W, Afzulpurkar N, et al. Design and simulation of MEMS based tuning fork micro-gyroscope [J]. Applied Mechanics and Materials, 2012, 110–116: 5036–5043
68. Riaz K, Bazaz S A, Saleem M M, et al. Design, damping estimation and experimental characterization of decoupled 3-DoF robust MEMS gyroscope [J]. Sensors and Actuators A: Physical, 2011, 172(2): 523–532
69. Cao Huiliang, Li Hongsheng, Shen Chong, et al. Bandwidth expanding system design of dual-mass silicon MEMS gyroscope [J]. Journal of Chinese Inertial Technology, 2016, 24(2): 218-223
70. Yang Bo. Research on silicon microgyro Measurement and Control Technology [D]. Doctoral dissertation of Southeast University, 2007
71. Cao H L , Zhang Y J , Yan J , et al. Mode-matching control system design of silicon MEMS gyroscope[J]. Journal of Chinese Inertial Technology, 2017
72. Yin Yong, Wang Shourong, Wang Cunchao, et al. Structural Design and Simulation of decoupled dual-mass Silicon microgyro [J]. Journal of Southeast University (Natural Science Edition), 2008, 38(5): 918–922
73. Cao Huiliang, Li Hongsheng, Shen Chong, et al. Silicon MEMS gyroscope bandwidth expanding based on bipole compensation technology [J]. Chinese Journal of Scientific Instrument, 2015, 36(11): 2427-2434
74. Cao H, Li H. Investigation of a vacuum packaged MEMS gyroscope architecture's temperature robustness [J]. International Journal of Applied Electromagnetics and Mechanics, 2013, 41: 495–506
75. Zhou W, Chen Z, Zhou B, et al. Identification of the coupling forces ratio in silicon vibratory MEMS gyroscopes [C]. Instrumentation and Measurement Technology Conference, 2010. 668–671
76. Cui J, Lin L, Guo Z, et al. Study on the quadrature error under Varying ambient pressure for a bulk micromachined Z-axis gyroscope [C]. Proc. 9th ICSICT, 2008. 2440–2443
77. Cui J, Guo Z, Zhao Q, et al. Force rebalance controller synthesis for a micromachined vibratory gyroscope based on sensitivity margin specifications [J]. Journal of Microelectromechanical Systems, 2011, 20(6): 1382–1394
78. Ding H, Yang Z, Yan G, et al. MEMS gyroscope control system using a band-pass continuous-time sigma-delta modulator [C]. IEEE Sensors 2010 Conference, 2010. 868–872
79. Cui J, Chi X Z, Ding H T, et al. Transient response and stability of the AGC-PI closed-loop controlled MEMS vibratory gyroscopes [J]. Journal of Micromechanics and Microengineering, 2009, 19: 125015
80. He C, Zhao Q, Cui J, et al. A research of the bandwidth of a mode-matching MEMS vibratory gyroscope [C]. Proc. Int. Conf. MEMS, 2012. 738–741
81. Li Jinming. Study on capacitive micromechanical gyro with high SNR [D]. Doctoral dissertation of North University of China, 2005
82. Cao H, Cai Q, Zhang Y, et al. Design, Fabrication and Experiment of a Decoupled Multi-Frame Vibration MEMS Gyroscope[J]. IEEE Sensors Journal, 2021, PP(99): 1-1
83. Chen Weiping, Chen Hong, Guo Yugang, et al. Two-stage Decoupling Mechanism Characteristics of a Fully symmetric micromechanical gyro [J]. Nanotechnology and Precision Engineering, 2009, 7(3): 239–244
84. Wang W, Lv X, Sun F. Design of micromachined vibratory gyroscope with 2 degree-of-freedom drive-mode and sense-mode [J]. IEEE Sensors Journal, 2012, 12(7): 2640–2464

85. Hou Z, Xiao D, Wu X. Effect of axial force on the performance of micromachined vibratory rate gyroscopes [J]. Sensor, 2011, 11: 296–309
86. Su J, Xiao D, Chen Z, et al. Dynamic force balancing for the sense mode of a silicon microgyroscope [J]. Measurement Science and Technology, 2013, 24: 095105
87. Xiao D, Su J, Chen Z, et al. Improvement of mechanical performance for vibratory micro-gyroscope based on sense mode closed-loop control[J]. J. Micro/Nanolith. MEMS MOEMS, 2013, 12(2): 023001
88. Su J, Xiao D, Chen Z, et al. Improvement of bias stability for a micromachined gyroscope based on dynamic electrical balancing of coupling stiffness [J]. J. Micro/Nanolith. MEMS MOEMS, 2013, 12(3): 033008
89. Chang H, Xie J, Fu Q, et al. Micromachined inertial measurement unit fabricated by a SOI process with selective roughening under structures [J]. Micro & Nano Letters, 2011, 6(7): 486–489
90. Feng R, Qiu A, Shi Q, et al. A research on temperature dependent characteristics of quality factor of silicon MEMS gyroscope [J]. Advanced Materials Research, 2010, 159: 399–405
91. Feng R, Qiu A, Shi Q, et al. A theoretical and experimental study on temperature dependent characteristics of silicon MEMS gyroscope drive mode[J]. Advanced Materials Research, 2012, 403–408: 4237–4243
92. Sun Jiaqing. Electric Compensation Method for Manufacturing Error of Silicon micro-vibration gyro [D]. Nanjing: Nanjing University of Science and Technology, 2008
93. Mo B, Zhou H, Zheng Q, et al. Quadrature error and offset error suppression methods for micro-gyroscopes [J]. Key Engineering Materials, 2012, 503: 174–178
94. Li H S, Cao H L, Ni Y F. Electrostatic stiffness nulling for quadrature error in decoupled dual-mass MEMS gyroscope [J]. J. Micro/Nanolith. MEMS MOEMS, 2014, 13(3): 033003
95. Chaumet B. Vibratory gyroscope balanced by an electrostatic device [P]. United States Patent, 20080282833A1
96. Handrich E. Method for measurement of rotation rates/accelerations using a rotation rate Coriolis gyro, as well as a Coriolis gyro which is suitable for this purpose [P]. United States Patent, 20070144254A1
97. Handrich E, Geiger W. Method for quadrature-bias compensation in a Coriolis gyro, as well as a Coriolis gyro which is suitable for this purpose [P]. United States Patent, 7481110B2
98. Lapadatu D, Blixhavn B, Holm R, et al. SAR500 – a high-precision high-stability butterfly gyroscope with north seeking capability[C]. IEEE/ION Position Location and Navigation Symposium, 2010. 6–13
99. Alper S E, Temiz Y, Akin T. A compact angular rate sensor system using a fully decoupled silicon-on-glass MEMS gyroscope [J]. Journal of Microelectromechanical Systems, 2008, 17 (6): 1418–1429
100. Ding H T, Liu X S, et al. A high-resolution silicon-on-glass Z axis gyroscope operating at atmospheric pressure [J]. IEEE Sensors Journal, 2010(10), 6: 1066–1074
101. Cui J, Guo Z Y, Zhao Q C, et al. An electrostatic force feedback approach for extending the bandwidth of MEMS vibratory gyroscope [J]. Key Engineering Materials, 2011, 483: 43–47
102. He C H, Zhao Q C, Liu Y X, et al. Closed loop control design for the sense mode of micromachined vibratory gyroscopes [J]. Science China Technological Sciences, 2013, 56(5): 1112–1118
103. Jia Fangxiu, Qiu Anping, Shi Qin, Su Yan. Design and Performance Test of silicon micro-vibratory gyroscope [J]. Optical precision engineering, 2013, 21(5): 1272–1281
104. Cao Huiliang. Dual-Mass Linear Vibration Silicon-Based MEMS Gyroscope [M]. National Defense Industry Press, 2020
105. Qiu Anping, Su Yan, Wang Shourong, et al. Effect of Residual Stress on the Performance of Z-axis silicon micromechanical vibratory gyroscope[J]. Journal of Mechanical Engineering, 2005, 41(6): 228–232
106. Mohd-Yasin F, Nagel D J, Korman C E. Noise in MEMS [J]. Measurement Science and Technology, 2010, 21: 012001

107. Mohd-Yasin F, Nagel D J, Ong D S, et al. Low frequency noise measurement and analysis of capacitive micro-accelerometers [J]. Microelectronic Engineering, 2007, 84: 1788–1791
108. Mohd-Yasin F, Korman C E, Nagel D J. Measurement of noise characteristics of MEMS accelerometers [J]. Solid-State Electronics, 2003, 47: 357–360
109. Mohd-Yasin F, Nagel D J, Ong D S, et al. Low frequency noise measurement and analysis of capacitive micro-accelerometers: temperature effect [J]. Jpn. J. Appl. Phys., 2008, 47: 5270–5273
110. Mohd-Yasin F, Zaiyadi N, Nagel D J, et al. Noise and reliability measurement of a three-axis micro-accelerometer [J]. Microelectronic Engineering, 2009, 86: 991–995
111. Mohammadi A, Yuce M R, Moheimani O R. A low-flicker-noise MEMS electrothermal displacement sensing technique [J]. Journal of Microelectromechanical System, 2012, 21: 1279–1281
112. Leland R P. Mechanical-thermal noise in MEMS gyroscopes[J]. IEEE Sensors Journal, 2005, 5: 493–500
113. Gabrielson T B. Mechanical-thermal noise in micromachined acoustic and vibration sensors [J]. IEEE Transactions on Electron Devices, 1993, 40: 903–909
114. Annovazzi-Lodi V, Merlo S. Mechanical-thermal noise in micromachined gyros [J]. Micro electronics Journal, 1999, 30: 1227–30
115. Kim D, M'Closkey R. Noise analysis of closed-loop vibratory rate gyros [C]. 2012 American Control Conference, 2012. 92–97
116. Liu Mei, Zhou Bailing. Mechanical Thermal Noise of silicon microgyro [J]. Chinese Journal of Scientific Instrument, 2006, 27 (6): 1163–1164
117. Shi Qin. Research on Several Key Technologies to Improve the Performance of Silicon micromechanical gyroscope [D]. Doctoral dissertation of Southeast University, 2005
118. Xia Guoming. Study on performance and stability of silicon microgyroscope based on digital Technology [D]. Doctoral dissertation of Southeast University, 2011
119. Hu Hao, DONG Jingxin, WAN Caixin, et al. Error Analysis of micro-accelerometer based on diode capacitance Detection [J]. Journal of Tsinghua University (Natural Science), 2008, 48(11): 1908–1910
120. Cao Huiliang, Li Hongsheng, WANG Shourong, et al. Structure Model and System Simulation of MEMS Gyroscope [J]. Chinese Journal of Inertial Technology, 2013, 21(4): 524–529
121. Wen Baiqian, Liu Jianye, Li Rongbing. System-level Modeling and Simulation of MEMS Gyroscope [J].Chinese Journal of Inertial Technology, 2007, 15(4): 485–487
122. Cao H L, Li H S. A novel temperature compensation method for a MEMS gyroscope oriented on a periphery circuit [J]. International Journal of Advanced Robotic Systems. 2013, 10(327): 1–10
123. Yang Jun, GAO Zhongyu, Zhang Rong, et al. Control technology for Structure Error of Micromechanical gyroscope [J]. Chinese Journal of Inertial Technology, 2007, 15(4): 488–493
124. Shi Qin, Qiu Anping, Su Yan, et al. Error analysis of silicon microgyro [J]. Journal of Sensing Technology, 2006, 19(5): 2812–2815
125. Painter C. Micromachined vibratory gyroscopes with imperfections [D]. University of California Irvine, 2005
126. Kim Y W, Yoo H H. Design of a vibrating MEMS gyroscope considering design variable uncertainties [J]. Journal of Mechanical Science and Technology, 2010, 24 (11): 2175–2180
127. Lv B, Liu X, Yang Z. Simulation of a novel lateral axis micromachined gyroscope in the presence of fabrication imperfections [J]. Microsyst Technol, 2008, 14: 711–718
128. Shi Qin, Qiu Anping, Su Yan, et al. Mechanical Coupling Error Analysis of silicon microgyroscope [J]. Optical precision engineering, 2008, 16(5): 894–898
129. Painter C C, Shkel A M. Identification of anisoelasticity for electrostatic 'trimming' of rate integrating gyroscope [C]. Proc. Smart Structures and Mater, 2001
130. Walther A, Blanc C L, Delorme N, et al. Bias contributions in a MEMS tuning fork gyroscope [J]. Journal of Microelectromechanical Systems, 2013, 22(2): 303–308
131. Weinberg M S, Kourepenis A. Error sources in in-plane silicon turning-fork MEMS gyroscopes [J]. Journal of Microelectromechanical Systems, 2006, 15(3): 479–491

132. Seeger J, Park M, Rastegar A, et al. Method and apparatus for electronic cancellation of quadrature error [P]. United States Patent, 7290435B2
133. Wen Baiqian, Liu Jianye, Li Rongbing. Orthogonal Error Analysis and Simulation of MEMS gyro [J]. Sensors and Microsystems, 2008, 27 (9): 82–88
134. Luo Bing, Zhang Hui, Wu Meiping. Orthogonal Error of silicon microgyro and its Influence on Signal Detection [J]. Chinese Journal of Inertial Technology, 2009, 17 (5): 604–613
135. Antonello R, Oboe R, Prandi L, et al. Open loop compensation of the quadrature error in MEMS vibrating gyroscopes [C]. IEEE 35th Annual Conference of Industrial Electronics, 2009. 4034–4037
136. Yeh B Y, Liang Y C. Modelling and compensation of quadrature error for silicon MEMS microgyroscope [C]. IEEE PEDS, 2001. 871–876
137. Liu Xue, Chen Zhihua, XIAO Dingbang, et al. Orthogonal Error Self-compensation method of vibratory micro-gyro [J]. Journal of Sensing Technology, 2012, 25(9): 1221–1225
138. Cao H, Li H, Kou Z, et al. Optimization and Experimentation of Dual-Mass MEMS Gyroscope Quadrature Error Correction Methods[J]. Sensors, 2016, 16(1): 71
139. Yang B, Zhou B, Wang S. Quadrature error and offset error suppression circuitry for silicon micro-gyroscope [J]. Journal of Southeast University (English Edition), 2008, 24(4): 487–491
140. Xia D, Wang S, Zhou B. A novel closed-loop vacuum silicon microgyroscope [J]. Journal of Southeast University (English Edition), 2009, 25(1): 63–67
141. Tatar E. Quadrature error compensation and its effects on the performance of fully decoupled MEMS gyroscopes [D]. Middle East Technical University, 2010
142. Cao Huiliang, Wang Yuliang, Shi Yunbo, et al. Optimization of quadrature error correction-method for silicon MEMS gyroscopes [J]. Optics and Precision Engineering, 2016, 24(1): 134-142
143. Wang XIAolei, Yang Cheng, Li Hongsheng. Analysis and Design of Orthogonal Error Correction System for Silicon microgyroscope [J]. Chinese Journal of Inertial Technology, 2013, 21(6): 822–827
144. Wang Pan, HUANG Libin, LI Hongsheng, et al. Study on closed-loop Orthogonal Correction of silicon microgyro [J]. Journal of Sensing Technology, 2013, 26(3): 357–360
145. Cui J, He C, Yang Z, et al. Virtual rate-table method for characterization of microgyroscopes [J]. IEEE Sensors Journal, 2012, 12(6): 2192–2198
146. Alshehri A, Kraft M, Gardonio P. Two-mass MEMS velocity sensor: internal feedback loop design [J]. IEEE Sensors Journal, 2013, 13(3): 1003–1011
147. Hu Shousong. Principle of Automatic Control (4th Edition)[M]. Beijing: Science Press, 2001
148. Wang Fuyong. A Correction Method for Eliminating Unsatisfied Complex Conjugate Poles [J].Journal of Soochow University (Engineering Edition), 2004, 24(6): 25–28
149. Cao Huiliang, Li Hongsheng, WANG Shourong, et al. Temperature Compensation of Silicon Micromechanical Gyroscope Measurement and Control Circuit [J]. Optical precision engineering, 2013, 21(12): 3118–3125
150. Xia Deqian, Weng Yifang. Automatic Control Theory (2nd Edition)[M]. Beijing: China Machine Press, 2004
151. Zeng S K. Reliability Design and Analysis [M]. Beijing: National Defense Industry Press, 2011
152. Cao Huiliang, Zhang Yingjie, et al. Pole-Zero Temperature Compensation Circuit Design and Experiment for Dual-Mass MEMS Gyroscope Bandwidth Expansion[J]. Mechatronics IEEE/ASME Transactions on, 2019
153. Cao H, Liu Y, Zhang Y, et al. Design and Experiment of Dual-Mass MEMS Gyroscope Sense Closed System Based on Bipole Compensation Method[J]. IEEE Access, 2019, 7: 49111-49124
154. Joint test group of micromechanical gyro. Test specification for micromechanical gyroscope [S]. Beijing: Micromechanical gyroscope Joint Test Group, 2010
155. IEEE Standards Board. IEEE standard specification format guide and test procedure for single-axis interferometric fiber optic gyros [S]. Gyro and Accelerometer Panel of the IEEE Aerospace and Electronic Systems Society, 1997

156. Yin Yong. Design Theory and Method of Dual-mass Silicon Micromechanical Gyroscope [D]. Doctoral dissertation of Southeast University, 2011
157. Yang Bo, Dai Bo, Liu Xiaojun, et al. The on-chip temperature compensation and temperature control research for the silicon micro-gyroscope [J]. Microsyst Technol, 2015, 21: 1061–1072
158. Cao Huiliang, Yang Bo, Xu Lu, Li Hongsheng, Wang Shourong. Design of MEMS Gyroscope Chip Temperature Control System [J]. Journal of Southeast University (Natural Science Edition), 2013, 43 (1): 55–59
159. Cao Huiliang. Research and Experiment on Electrostatic Compensation and Control Technology of Silicon Micromechanical Gyroscope [D]. Doctoral dissertation of Southeast University, 2014
160. Cao Huiliang, Li Hongsheng, Liu Jun et al. An improved interface and noise analysis of a turning fork microgyroscope structure [J]. Mechanical Systems and Signal Processing, 2016, 70–71: 1209–1220
161. Chang Honglong, Xie Jianbin, Fu Qianyan, et al. Micromachined inertial measurement unit fabricated by a SOI process with selective roughening under structures [J]. Micro & Nano Letters, 2011, 6(7): 486-489
162. Chen Shuling. Study on temperature stability of silicon microgyro [D]. Master's thesis of Southeast University, 2010
163. Sheng Xia, Wang Shourong, Chen Shuling, et al. Realization of PID Algorithm for Temperature Control System of Silicon Micromechanical Gyroscope [J]. Measurement and Control Technology, 2010, 29(8): 4-12
164. Xia Dunzhu, Chen Shuling, Wang Shourong, et al. Microgyroscope temperature effects and compensation-control methods [J]. Sensors, 2009, 9: 8349-8376
165. Wang Xiaolei. Research and Experiment on Closed-loop Detection and Orthogonal Correction technology of Silicon Microgyro [D]. Doctoral dissertation of Southeast University, 2014